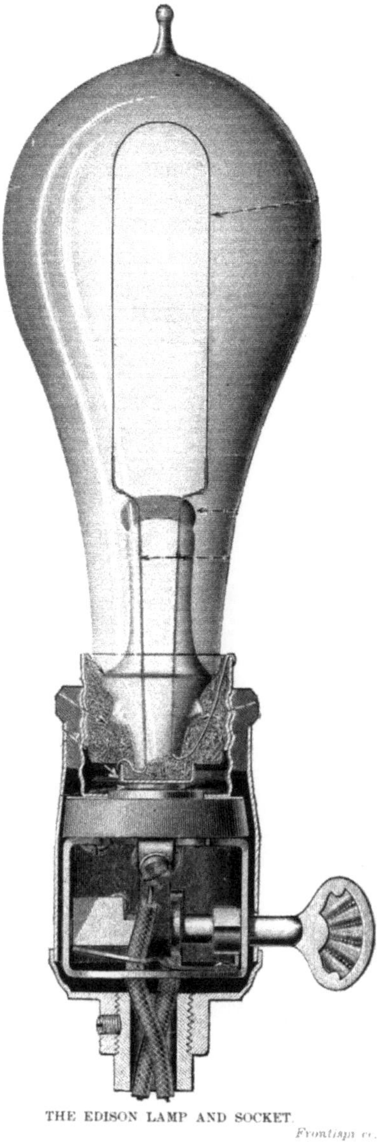

THE EDISON LAMP AND SOCKET.

Frontispiece.

INCANDESCENT

ELECTRIC LIGHTING.

A Practical Description of the Edison System.

BY

L. H. Latimer.

TO WHICH IS ADDED THE

DESIGN AND OPERATION OF INCANDESCENT STATIONS.

By C. J. Field.

AND A PAPER ON

THE MAXIMUM EFFICIENCY OF INCANDESCENT LAMPS.

By John W. Howell.

AFRIWARE ™

Afriware Books, Co

ISBN:979-8-9851194-3-5

Originally printed: 1880

Reprinted: 2025

Published and Distributed By:
Afriware Books, Co
Books, Gifts and Cultural Events
137 N. Oak Park Ave.,
Suite 207
Oak Park, IL 60301,
708-223-8081

ISBN: 979-8-9851194-3-5

INCANDESCENT

ELECTRIC LIGHTING.

BY

L. H. LATIMER.

Editor's note

We are honored beyond measure to reprint "Incandescent Electric Lighting", a masterpiece that stands as a breathtaking testament to Black excellence. In an era when the worst was expected of Black people, Lewis Latimer not only defied expectations—he redefined them. As the only person of African ancestry among Edison's elite circle called, "Edison Pioneers", he made sure his presence was undeniable. Every technical illustration in this book was drawn by his own meticulous hand, a testament to the depth of his knowledge and the precision of his craft. To create such detailed images, Latimer had to possess an intimate understanding of electric lighting, down to its very core. At the time of its publication, this book was regarded as the ultimate reference on electric lighting—practically a bible for the industry. Latimer's contributions weren't just theoretical; they were revolutionary. His carbon filament extended the life of the electric light bulb from a mere day to over a month, transforming a fragile invention into a practical, world-changing technology. He worked alongside many great pioneers of the era—Alexander Graham Bell, Thomas Edison, Maxim, Weston—yet history too often fails to give him his due. He is an inductee of the National Inventors Hall of Fame for his work on electric filament manufacturing techniques. But we will not let the Light of his name be dimmed. This book stands as undeniable proof of his genius, his perseverance, and his invaluable role in shaping the modern world.

STILL. WE. RISE.

We are duty-bound to share the stories of brilliance, achievement, and resilience with all who will listen. Lewis H. Latimer's legacy deserves nothing less.

Nzingha Nommo
Owner/Founder
Afriware Books, Co

Ebon Venus

Let others boast of maidens fair,
Of eyes of blue and golden hair;
My heart like needles ever true
Turns to the maid of ebon hue.

I love her form of matchless grace, The dark brown beauty of her face, Her lips that speak of love's delight, Her eyes that gleam as stars at night.

O'er marble Venus let them rage,
Who set the fashions of the age;
Each to his taste, but as for me,
My Venus shall be ebony.

-Lewis H. Latimer, "Black Pioneers of Science and Invention" by Louis Haber

INCANDESCENT

ELECTRIC LIGHTING.

At the time when the first edition of this little volume issued from the press, Electric Lighting by Incandescence, had barely got beyond the stage of a laboratory experiment, and the possibility of lighting large districts from a central station, as successfully as with gas, was doubted by men who now stand high in electrical circles. So great indeed has been the change in this art since that time, that we are rapidly becoming possessed of the indifference consequent upon familiarity, and take the introduction of the electric light into our factories, business places, and dwellings, as a matter of course.

There is however one phase of this industry, which is rapidly placing—even a general knowledge, of the means by which the electric current is generated and distributed beyond the reach of all but those actually connected with it as a business, or the student who proposes to follow it as a profession; this is the tendency to centralize electric lighting and to substitute large central stations, where the machinery for generating the current is located and whence the wires are run for distributing the same, for the numerous small plants now scattered over the country.

While these central stations cheapen the production of the light, and bring it within the reach of those who otherwise could not afford it, it does away with the large number of isolated plants, which formerly afforded the curious an opportunity to inspect the generation, distribution and utilization in light, of this form of energy.

While the opportunities to become informed upon this subject are rapidly growing less, the electric light as a factor in our civilization, is becoming daily of more importance,

Millions of capital are being invested in its production, and it is being introduced throughout the world as rapidly as human activity, supplemented by all the agencies of an advanced civilization, can accompiish this end.

It is to meet the want among the intelligent laity, that is now—and will be later on more keenly—felt, for a general knowledge of this subject, superficial perhaps, but yet connected and logically arranged, that the publishers have decided to re-issue this number of their Science Series, with the matter contained therein, thoroughly revised and brought down to date; and trust that the same generous appreciation which attended their first effort, will justify them in this attempt to give to the public a popular exposition of the art of electric lighting by incandescence.

While a system of electric lighting, includes a large number of devices, the greater portion of these are simply designed to control or utilize the current produced by the generator. This latter then is practically the alpha and omega of all systems, for from it goes forth, and to it returns, the energy which in different parts of the circuit, take the form of light, through the instrumentality of the carbon filament enclosed in the glass bulb of the lamp. The generator acts somewhat like a pump, having a reservoir within it, and which, when in motion, forces the water from the reservoir, through a long loop of pipe, both ends of which terminate in itself.

It is thus, that the circuit may be said to commence and end at the generator, which transforms the mechanical energy imparted to it into electrical energy, and through the medium of electricity, produces effects that would be otherwise unobtainable.

While the laws, upon which the construction and operation of the electric generator depends, are complex, it is yet possible to understand its parts, their relation to, and action upon each other, without entering deeply into their intricacies.

We are all familiar with the common horse-shoe magnet, and have used it as a toy, if not otherwise.

While watching how readily a needle, tack, or other small object of iron or steel, would follow its movements, we have wondered how this motion was produced without the objects coming into actual contact with the magnet, and have rested content with the explanation that the phenomena before us, was the result of a force called magnetism, which resided in the magnet.

This force which stretched forth its invisible fingers to move the tiny needle, is that upon which depends the action of the electric generator; the magnet however, differing in size, and also in the method by which its magnetic powers are excited, as will be explained hereafter. For the present let us confine ourselves to the horse-shoe magnet with which we are so familiar.

If we take such a magnet, and wave it

to and fro above a needle resting upon the smooth top of a table, we will find that the needle will follow the magnet energetically or otherwise, according to the proximity of the magnet; that is to say, the farther the magnet is away the less its power over the needle.

It may be well to state here, that, when speaking of the action of the magnet upon any object, we wish it to be understood that the object is presented to the ends of the magnet, as it is in its ends that its greatest strength lies. The relative conditions of the different parts of a straight-bar magnet are shown in Fig. 1, the brush-like lines at the ends giving a clear idea by their number and length, of the different degrees of magnetism existing at various points in the magnet;

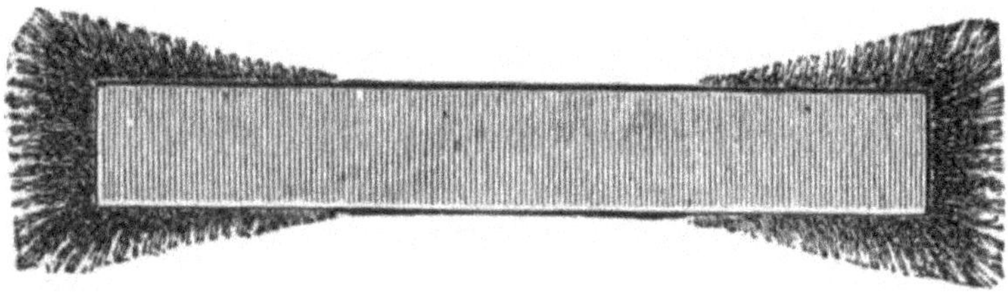

FIG. 1.

these conditions would remain practically the same were the magnet bent in horseshoe form.

If we substitute for the needle a loop of copper wire we shall, by the motion before described, induce in the copper wire a current of electricity, and that without actual contact between the magnet and the wire.

Should you hold a horsehoe-magnet a slight distance above a long bar of iron, and then move it forward parallel thereto, you would find that it offered resistance to being moved thus, and that you would be forced to put forth some strength, both to keep it in motion as well as to prevent it being drawn down into contact with the iron; in fact, the sensation would be much the same as that experienced in moving a brush across the surface of a liquid, oil for instance, with the bristles just touching its surface.

These invisible bristle-like projections produce the current of electricity in the wire when the latter cuts through them.

To tell why such results follow the movements we have described we leave to higher authorities; our purpose is to deal with the facts, believing this sufficient for the purpose we have in view.

If you will glance at figure 2 you will see there shown a loop of wire such as has been referred to, mounted upon a shaft having a handle for the purpose of

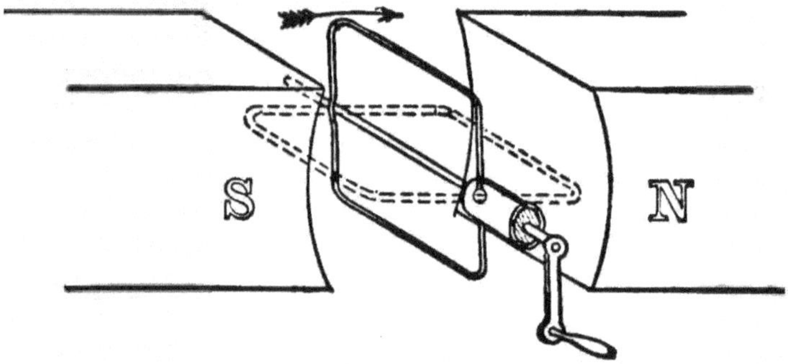

FIG. 2.

revolving the loop rapidly between the ends of a magnet. This wire is covered with a substance which prevents the electricity passing from it otherwise than through the large pieces of metal, which are arranged about a cylindrical piece of non-conducting material, as a block of wood, in such a manner as not to touch each other; this latter being shown in the cut located near the handle on the forward end of the shaft. Figure 3 shows an end view of this insulating device.

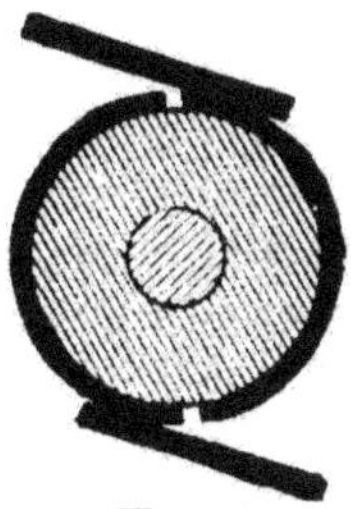

FIG. 3.

If the shaft which supports the loop is rapidly revolved, a current of electricity will be set up in the wire loop by bridging over the space at the ends of the wire between the metal plates with a piece of metal or another loop of copper wire; copper always being used to conduct electricity where it can be, for the reason that it is the best practical conductor of electricity.

Such a device as this would be of little service in producing the large currents needed for the electric light, both because there would be an insufficient amount of wire and, further, because such magnets as have been described—called permanent magnets—cannot be made sufficiently strong for the commercial generation of great currents.

We therefore have to resort to methods which, while they differ somewhat from this arrangement in detail, are practically the same.

In the first place we must have a large number of these wire loops, and as the wire is soft and easily bent, we must wind it about a core so as to hold it rigidly in place, as it has to be revolved very rapidly. Again, the ends of each wire loop must be brought out and attached to plates or blocks of metal, arranged upon some kind of non-conducting substance at one end of the shaft upon which the whole will turn. Sometimes the ends of the loops are attached to each other in such a manner as to combine all the loops as though they were one wire, and then separate pieces of wire are attached to them where they join each other, and these short wires terminate in plates at the end of the shaft as before described. As each loop of wire (which may be made of one or more coils) comes into the space between the ends of the magnet, a current is set up in it; this current increasing or subsiding, according to the position of the loop with relation to the ends of the magnet. As the current is generated, it flows toward the pieces of metal arranged on the end of the shaft, and from them into what are called brushes, which are flat pieces of metal that rest against those arranged on the end of the shaft so as to have intimate contact with them without retarding their rotation.

This arrangement is clearly shown in the cut, Fig. 4, in which *a* is a shaft, to which is fixed a drum or core *b*, wound with a number of coils of insulated wire *c*; that is, wire which is carefully covered with a material which practically prevents the passage of electricity through it. These wires are connected together at the front end of the drum, and are also secured to curved pieces of metal *d*, which for the sake of clearness, are shown with that end of the shaft removed, and as also having the brushes *e e* resting against them.

These plates, *d d*, vary in number

according to the number of coils of wire wound about the drum, and when arranged in place and secured to a drum or core of insulating material,—that is material which will not conduct electri-

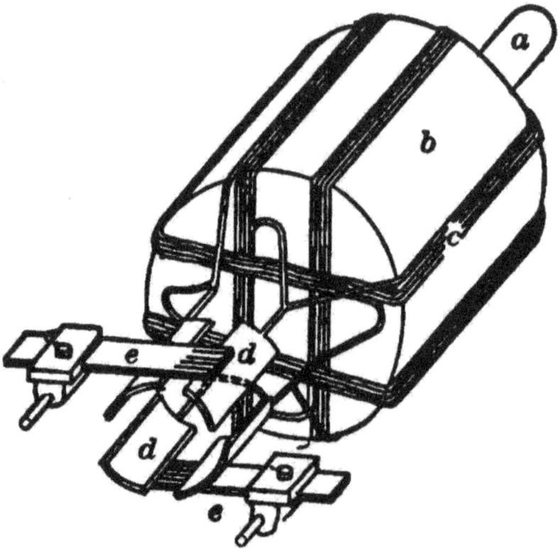

Fig. 4.

city, — are called the "commutator;" their office being to conduct the impulses of current from each coil as it comes into proper position with relation to the ends of the magnet, into the brushes, from which it passes out through the line wires to the lamps or other electrical devices.

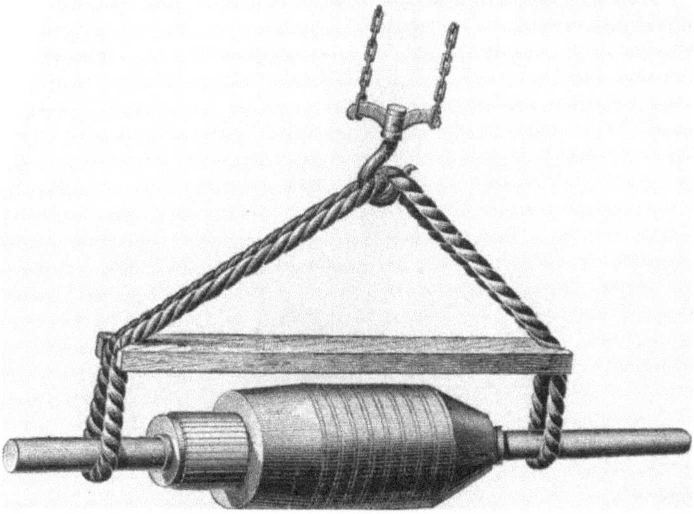

(To face page 19.)

The drum with the wire wound about it is called the "Armature."

In the practical machine used in electric lighting, the core of the armature is made of thin discs of soft sheet iron; these discs are prevented from touching the shaft by the interposition of insulating material, and have insulating material, arranged between them so as to prevent their touching each other, they are however so rigidly secured to the shaft as to be in no danger of becoming loose when the shaft is rapidly rotated. Fig. 5 shows a complete armature, with commutator attached ready to be placed in position in the generator.

When the parts of a generator are properly connected to each other, as well as to the outside lamp circuit, the current generated in the armature, is augmented by the presence of the iron core, which attracts the lines of force extending out from the field pieces, and so concentrates them as to cause the wire wound about it to cut through the greatest possible number of these lines in a given time.

And here it becomes necessary to lay aside the permanent horseshoe-magnet, which has served us so well up to this point in our explanation, and substitute in its place the practical electro-magnet.

If we retain the shape of the horseshoe (see fig. 6), and make our magnet of soft iron—instead of steel, the material of which the permanent horseshoe-magnets are made—we shall have a device which will need something to charge it with magnetism whenever it is brought into use; this something is the electric current, and we supply it to the magnet by winding one or more coils of insulated wire about the legs of the magnet near their ends.

It is necessary that there should be a slight trace of magnetism in these electro-magnets, in order that the electric current may have something to act upon to assist it in creating the necessary magnetic condition in the generator; this want is supplied by what is called residual magnetism, the name given to the small

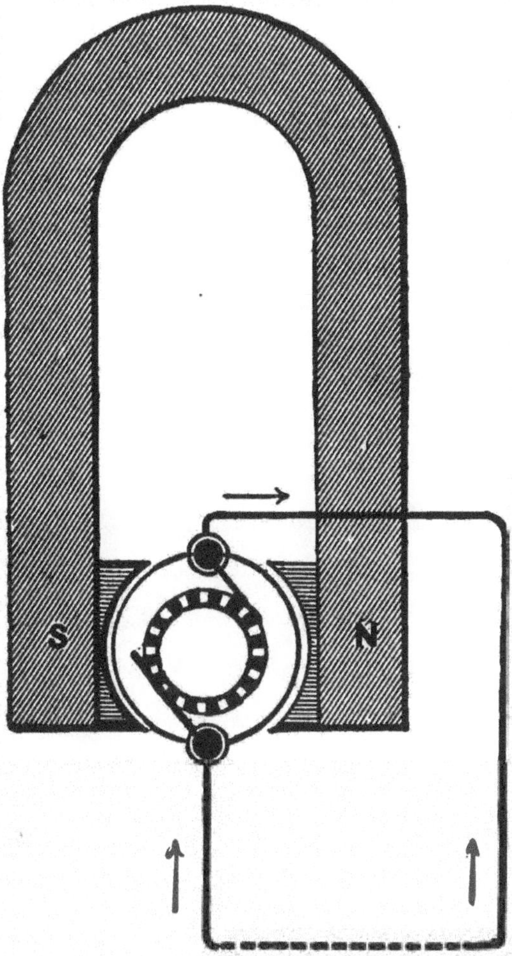

Fig. 6.

quantity of magnetism which always resides in large masses of iron or steel, or remains there after it has once been excited by an electric current.

These horseshoe-magnets are called the "Field Magnets" of a generator, because, between their ends where the armature is located is what is known as the field of force, taking its name from the lines of force, which is the technical term used for designating the power eminating from the poles of a magnet; and finally, in the practical generator the magnet ceases to assume the form of a horseshoe.

In Fig. 7 is shown a generator of the type now commonly used for supplying the current for the incandescent electric light. The ring having the alternate black and white spaces, is the commutator, which you will remember is secured to the same shaft as the armature and is practically the same. The large circle about the commutator is intended to represent the outer surface of the armature, which runs very close to the field pieces, N S, as these enlarged ends of the magnet are called. The heavy lines resting against the commutator are the brushes from which the wires leading to and from the lamps go forth.

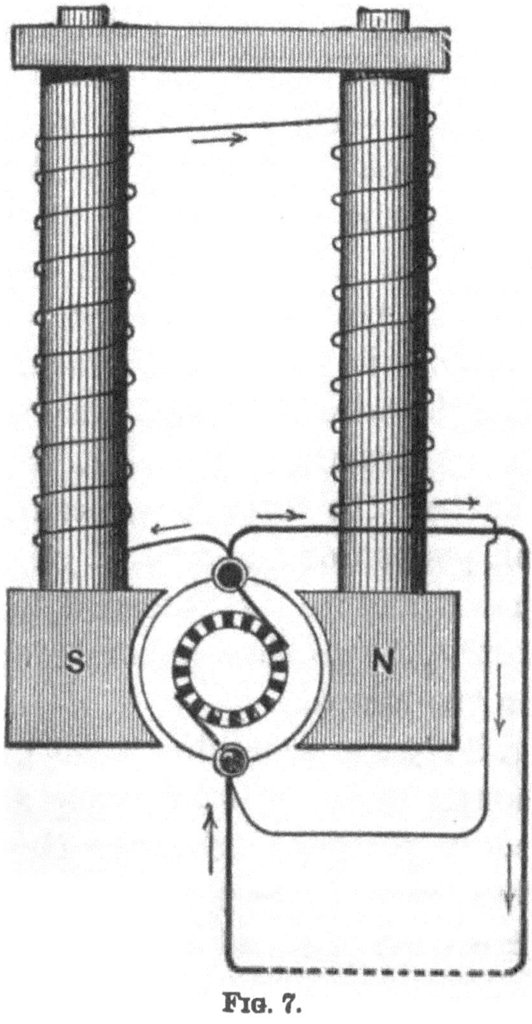

Fig. 7.

You will notice that from the upper brush, two wires are led; the larger of these, is supposed to be the wire that carries the current to the lamps, the location of which is shown by the dotted portion of the heavy line ; the direction of the current in the same being indicated by the small arrows as proceeding from the upper to the lower brush.

The fine wire passing from the upper to the lower brush is shown as being coiled about both legs of the magnet, and terminating in the lower brush. The office of this wire is to magnetize the iron in the electro-magnet, and it is made smaller than the lamp wire, because only a small portion of the current generated in the armature is intended to flow through it, the greater portion being utilized in the lamps.

Fig. 8 gives a clear illustration of this, the lamps being shown as they are arranged in practical lighting. Here also is shown, what is termed a resistance box, which is a box filled with a large number of coils of wire, made of material that will not readily conduct electricity, and so arranged as to be capable of being brought one after another into circuit with the fine wire passing around the field-magnets, so as to regulate the amount of current which passes about

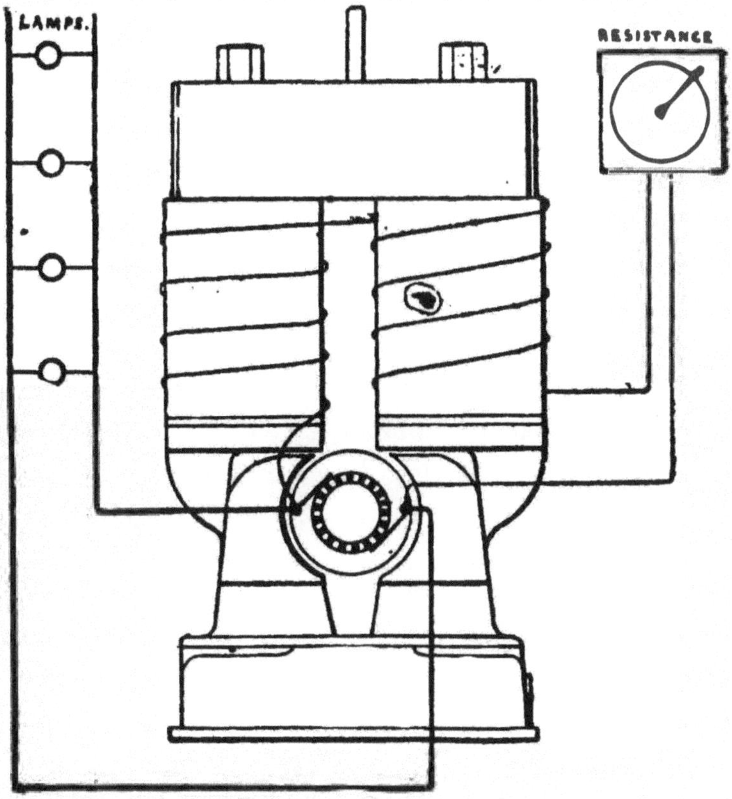

FIG. 8.

them, thus controlling the amount of current generated.

The form of generator shown in this

figure, is that used in the Edison system of incandescent electric lighting.

The pole pieces rest upon blocks of zinc, which do not interfere with their magnetized condition; and this zinc is in turn supported upon an iron base, the whole being secured together so as to form a neat and compact machine as shown in Fig. 9.

Before proceeding to discuss the means and devices for utilizing the electric current, let us be quite sure that we fully understand the several parts of the generator, or Dynamo, as it is commonly called, their individual purpose, action upon, and relation to each other. First are the electro-magnets known as "field-magnets," with their ends or "poles" enlarged by the "field-pieces," between which is the space known as the "field of force" or "magnetic-field," in which revolves the "armature," having its several "coils" of wire connected to the "strips" of the "commutator," these strips having intimate contact with the "brushes" resting upon them, which

FIG. 9.—Edison Dynamo Machine. (*To face page 26.*)

latter are connected together by the "line wires," the outgoing current wire generally passing forth from the upper brush to the locality to be lighted, from which the wire for the incoming current returns to be secured to the under brush, or what is equivalent to that, the line wires are secured in "binding posts" connected directly to the brushes. We thus have a "closed circuit," as it is generally termed, consisting of the wire coiled upon the armature, the strips or plates upon the commutator, the brushes, and the wire leading from and returning to them, between, or in which the lamps are arranged; so that the current of electricity generated in the armature, passes through all of these, and returns again to the armature, thus completely traversing what is known as the "lamp circuit."

We also have a line of finer wire passing from one of the brushes, coiling about the field-magnets, and returning to the other brush; this is known as the "field circuit."

The lamp circuit is so named, because

in it are located the incandescent lamps.

The field circuit takes its name from its office of furnishing the necessary current for the excitation of the field-magnets.

Having located and arranged all the parts of the generator, let us proceed to note how they act upon each other.

Attached to the shaft of the armature, is an iron wheel or pulley, about which passes the belt from a steam engine. When the engine is started, this belt transmits its motion to the pulley, rapidly revolving it and the armature upon the shaft with it.

As the armature revolves, the wire round about it is acted upon by the residual magnetism in the field pieces; this excites a slight current in the armature, a part of which passes out to the lamp circuit, the remainder passing through the field circuit and increasing the magnetism in the fields, which act somewhat more strongly upon the armature, inincreasing and strengthening the current passing from it, which in turn acts upon the magnets, and thus each aids the other, until the magnets have reached their full power, and we have the generator working at its greatest capacity; the major portion of the current flowing out to the lamps, while the remainder keeps the field magnets charged with the magnetism necessary for the generation of the current.

THE LAMP.

In Fig. 10 and Fig. 10*a* is shown an incandescent lamp, both in and out of its receptacle or socket. The lamp is formed of an elongated glass bulb, into the bottom of which passes a hollow glass stem; this stem is closed at its top and has imbedded in it small conducting wires of platinum which support a loop-like thread or filament of carbon or charcoal, made from the fibre of bamboo. The bottom of the tube is melted or sealed to the glass bulb so as to make the latter air tight at that point, and after the air contained in it is exhausted through a tube of glass attached to its top and connected with an air pump, that end is also hermetically sealed, by melting the glass at that point into a tiny glass knob, as shown in the cut.

Attention is then given to the bottom of the lamp; about which is moulded a

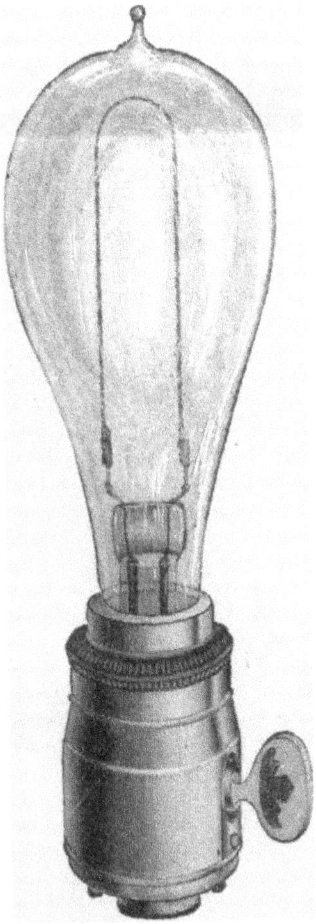

FIG. 10.

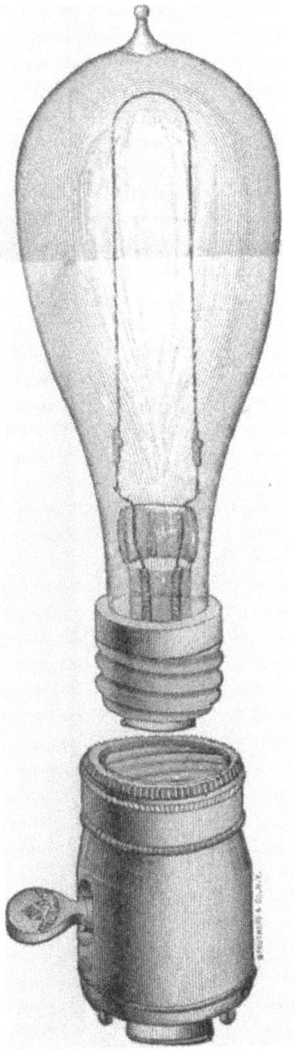

FIG 10a.

(To face page 30.)

block of plaster held in place by spurs or projections on the bottom of the tube which supports the carbon filament.

About the outside of this plaster base, is arranged a thin shell of metal pressed into the form of a screw, and to which is attached the end of a small copper wire, the other end of this wire being melted or soldered to the free end of one of the bits of platinum supporting the carbon filament ; the other bit of platinum being in a like manner connected by a piece of copper wire to a metal button on the bottom of the plaster base of the lamp.

Thus to follow the circuit of the completed lamp, we could start at the metal screw shell on the outside of the plaster base, thence through the copper wire attached to it to the small platinum wire embedded in the top of the glass tube, through the carbon filament secured to it, to the opposite platinum wire and down the copper wire secured thereto, terminating at the metal button on the bottom of the plaster base, which is insulated from the screw shell by the plaster of Paris, a non-conductor of electricity.

This is the course the electric current would take when the lamp is in operation.

In the socket is also arranged a screw shell, to receive that on the base of the lamp; while the button on the bottom of the lamp base, comes into intimate contact with a similar button in the lamp socket.

The socket has upon its bottom a threaded piece by which it may be arranged upon a gas fixture chandelier, or regular electric light fixture; and when thus arranged, has leading into it, two wires,—one permanently attached to a metal piece connected directly with the screw shell of the socket, while the other is attached to a metal piece which can be brought into circuit with the button in the socket, by means of a metallic connection carried on the shaft of a key arranged to rotate in the socket.

By rotating this key, shown projecting from the socket, the connection between this latter wire and the button of the socket can be either made or broken, thus enabling the lamp to be turned on and off at will.

If the electric current can be forced through a substance that is a poor conductor, it will create a degree of heat in that substance, which will be greater or less according to the quantity of electricity forced through it.

Upon this principle of the heating effect of the electric current, is based the operation of the incandescent lamp just described. While the copper and platinum wires readily conduct the current, the carbon filament offers a great deal of resistance to its passage, and for this reason becomes very hot, in fact is raised to white heat or incandescence, which gives its name to the lamp. You doubtless wonder why this thread of charcoal is not immediately consumed when in this state, but this is readily accounted for when you remember, that without the oxygen of the air, there can be no combustion, and that every possible trace of air has been removed from the bulb and it so thoroughly sealed up as to prevent the admission of the air about it; and yet the lamp does not last for ever, for the reason that the action of the current upon the carbon has a tendency to divide up its particles and transfer them from one point to another so that, sooner or later, the filament gives way at some point. Yet most of these lamps are guaranteed to last a thousand hours, and this at from four to six hours a day gives the lamp a life of several months.

Although electricity, like the air around us, seems very impalpable, appealing to so few of the senses, it is yet capable of being measured, for in order to run the lamps economically, we must give each of them only its due measure of the electric current passing over the wires.

The current which flows through each lamp is measured in "Ampères" by an "Ampère Meter," and the pressure which forces it through against the resistance of the carbon filament is measured in "Volts," by a "Volt Meter," so that a lamp to give a light equal to an ordinary gas burner is known to consume a certain amount of electric energy, and we are thus enabled to determine what it costs to produce light, or to burn each lamp for a given number of hours.

In addition to these instruments we have a device which is placed in each house to show how much current passes to the lamps therein; This is the "Electric Meter."

There are a number of methods of arranging the incandescent lamps with relation to the wires leading from the generator. What is termed the two wire system is shown in Fig. 8, and is the one generally employed in what is known as isolated lighting; that is, in cases, where the machinery for generating the current, is located at the place where the light is to be used; as in factories, large hotels, and like places not located in a district lighted from a central station.

When large districts of a city are lighted from a central station, what is termed the three-wire system, is generally employed, as it has economical advantages, which are not possessed by any other method that has been in practical operation up to the present time for commercial lighting on a large scale.

In the two-wire system, the lamps are arranged between the two large conductors, in what is called multiple arc or parallel; that is they are arranged like the rounds of a ladder between the main conductors.

This arrangement leaves each lamp independent of all others, so that any one may be turned on or off without interfering with those that remain burning. In order, however, that the lamps farthest removed from the generator shall have as much current as those nearest to it, the main conductor must of necessity be so large as to present practically no resistance to the passage of the current; for this reason the two-wire system cannot well be used, under conditions requiring the current to be led a long distance to the point where the lights are to be used, as the cost of the large copper conducting mains becomes so great as to seriously interfere with the economical introduction of the system upon a commercial basis. It was to obviate this defect, that the three-wire system was invented by Mr. Thomas A. Edison.

In practical electric lighting, the number of lamps receiving their current from a common central station, often runs up into thousands; and, as it is neither practicable nor desirable, to run these lamps all from one generator, a number of generators have to be combined together to furnish the necessary current, each generator having a capacity of a given number of lamps. These being the conditions under which lighting must be done, we will endeavor to make clear by comparison the advantages secured by the employment of the threei wire system. In. Fig. 11 A represents two dynamos having a capacity of five lamps each. Supposing these lamps to require one ampère of current each, and a total electro-motive force of pressure of ninety volts at the lamps. If these lamps were, say, five hundred feet away from the generator, we might in order to save expense in wire, use a size so

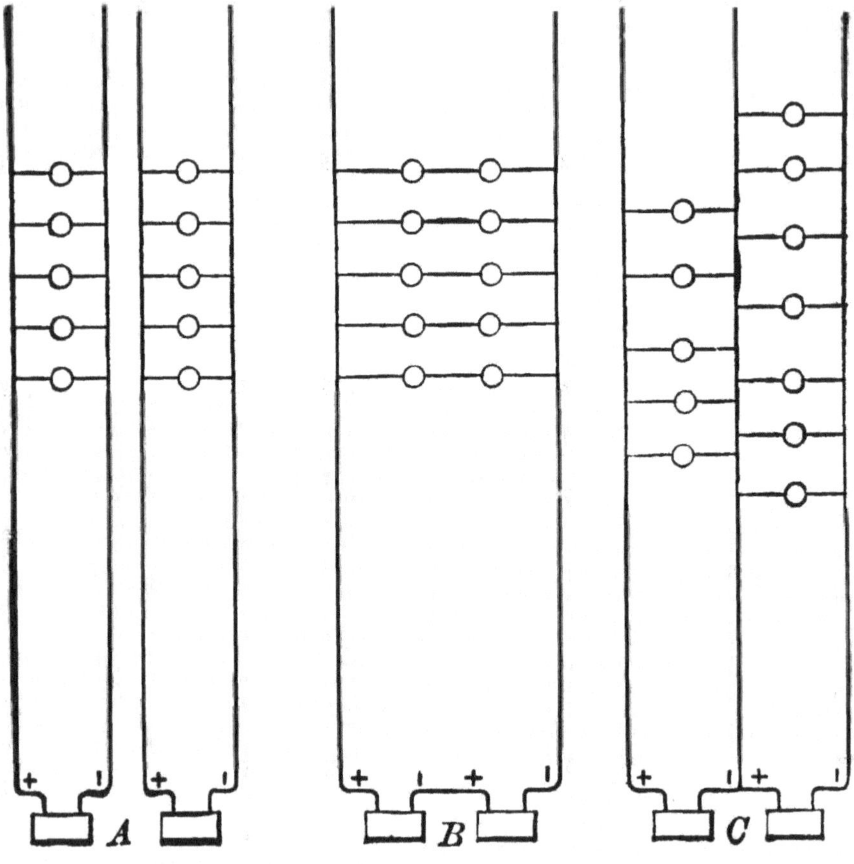

FIG. 11.

small as to entail a loss of ten per cent. of the total energy in the wires; that is, ten per cent. of the energy given out by the generator would be consumed in overcoming the resistance of the wire. This loss would amount to ten volts, it being much cheaper to waste this amount of energy in the wires, where great distances have to be traversed, than to furnish wires large enough to present no resistance to the passage of the current.

This being the case, the two dynamos shown at A, would be furnishing each 100 volts and 5 ampères, and would be using conducting wires or "Mains," large enough to carry this current. Should we now connect these ten lamps two and two in "Series," that is, two on the same wire as shown at B, between the outside wires of the two dynamos, and then connect the dynamos together by a short wire, we should be enabled to dispense with two of the mains leading from the generators to the lamps.

As the lamps were first arranged, it was only necessary to have a pressure of ninety volts, in this arrangement, however, two lamps being coupled together in series, their resistance is doubled, and consequently the electro-motive force to overcome that resistance must also be doubled; and we must now have 180 volts at the lamps, but not so with the current. The lamps being now arranged two on the same wire, the current which passes through one will pass through both; and now instead of having one ampère to each lamp, we have one ampère to each two lamps, or five ampères for the ten lamps.

This is just one half of the current used in the first case, as illustrated at A, we have therefore, by the removal of the two conductors, reduced our wires one half, and by the change in the arrangement of the lamps, have reduced our current one half. Since this arrangement, while an excellent one, has the disadvantage that the two lamps coupled together are dependent upon each other, and neither can be extinguished, without cutting off the supply of current from the other; to obviate this defect we have a third wire extending from the junction of the two dynamos, and passing between the lamps, as shown at C. This is termed the neutral wire, and serves to conduct the current to either lamp of a couple when the other is turned out. Supposing the neutral wire to be of the same size as the other two wires, we then have in the three-wire system, three wires, each one half the size of any one of the four wires used in the two-wire system, thus we have for the same number of lamps but three eighths of the original amount of wire.

In actual practice this is further reduced by making the neutral wire one half the size of the other two; thus giving us a total of five sixteenths only of the wire used in the arrangement shown at A.

Fig. 12 shows very clearly the arrangement of dynamos, wires, and lamps, in the three-wire system. A and B are the dynamos; the positive wire *r* of A and negative wire *t* of B, forming the positive and negative wires of the system, while the negative wire of A, and the positive wire of B, unite at *s* to the neutral wire of the system; *c*, *g*, *k*, and *m* show the lamps as practically arranged between the wires in the usual manner, while *h* and *i* show lamps arranged between the outside wires, as they may be, and frequently are, for certain specific purposes, and *e* and *f* illustrates the method of leading

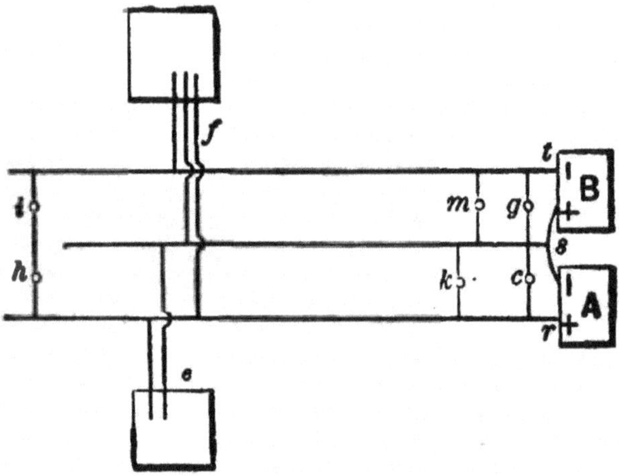

FIG. 12.

wires from the street mains into houses which are to be lighted, as is the practice where this system is to be used on a large scale.

The two generators shown coupled together form the unit which is multiplied in creating a central station. Of course the dynamos vary in size and capacity;

but whatever be the size of the station it is formed of two or more of these units, that is, of two or more couples of generators, each couple being composed of two dynamos as near alike as it is possible to have them, and run from one and the same engine. It will be seen that by this arrangement a station is secured against a general break-down; for the disablement of any one unit will in no way prevent the operation of the others, and as a station rarely runs at its full capacity there is always machinery enough at hand to supply the current cut-off by the failure of any unit to operate in a satisfactory manner.

It is important in operating the three-wire system that the number of lamps on either side of the middle or neutral wire should be the same. When this condition exists, the neutral wire has no current flowing through it and the system is said to be balanced; but as the lights are turned on and off at will, by the parties using them, it is obvious that the lights on different sides of the system will frequently vary, and consequently often throw the system out of balance. To counteract this influence certain devices known as "Equalizers" are introduced into the system.

These equalizers consist of boxes containing coils of wire which offer a great deal of resistance to the passage of the current, and are so arranged that one coil after another may be brought into or out of the circuit, by simply rotating a handle placed on the outside of the box.

The general arrangement of this device is shown in Fig. 13, where DD are dynamos, and RR the equalizers.

When a number of lamps sufficient to destroy the equilibrium of the system are turned in or out of either side of the system, a proportionate amount of resistance is thrown in or out of the proper side by means of the regulators so as to restore the balance.

Where large districts of a city are to be lighted it is important that the pressure at all points of the circuit should be uniform, in order that all lamps may give the same amount of light. To accomplish this by running conductors directly from the generators, would in-

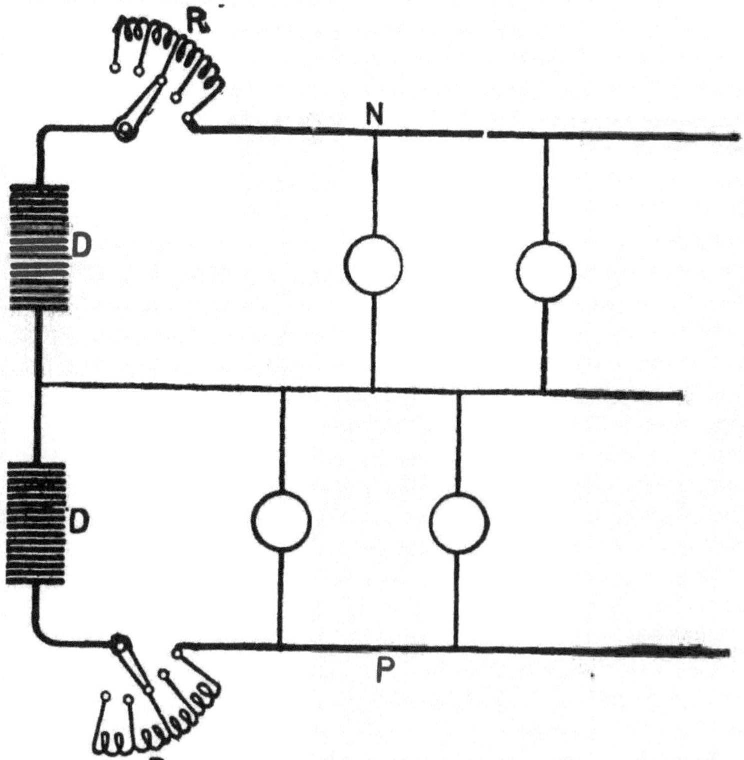

FIG. 13.

volve the use of wires so large as to practically debar their being used on account of the great cost of the copper employed in their manufacture.

In view of these conditions it is the practice to lay the "Mains," which are to supply the current to the lamps, only through the streets where the lamps are to be used. This is shown in Fig. 14 which represents the map of a section of a city district.

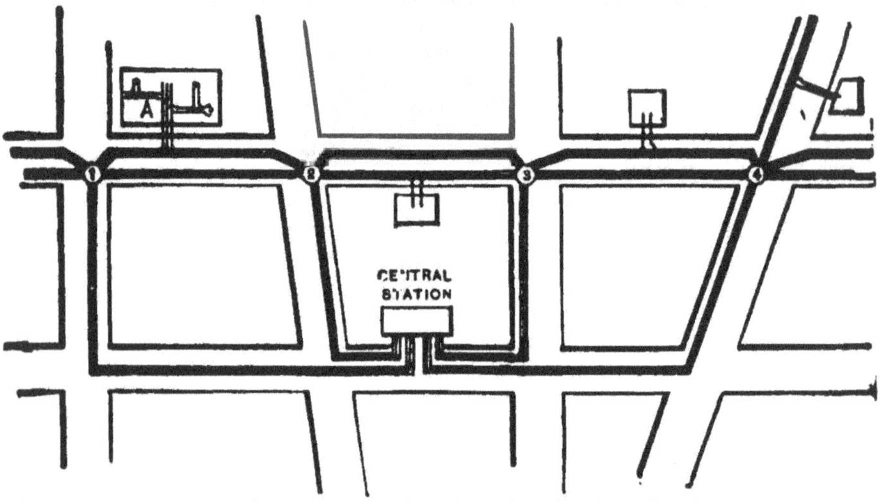

FIG. 14.

The mains are laid underground on both sides of the street; and are joined gether in "Man-Holes," or "Junction-Boxes" at the corners.

Into these junction-boxes are brought what are termed the "Feeders" which convey the current from the station to the mains. The feeders are brought in at such points as are best arranged for distributing the current equally throughout the mains.

In Fig. 14, the junction-boxes are numbered 1, 2, 3, 4, and the lines running from them to the station are the feeders. In addition to the three feeders, there are three "Pressure-Wires" returning to the station, these latter being connected to what are termed "Pressure Indicators," located in the station in plain view of the operators.

These pressure indicators show the pressure at the ends of the feeders in the junction-boxes where they are connected to the mains, and warn the attendant when the pressure in any part of the system is either higher or lower than it should be.

Where wires are to be laid underground they are first carefully wrapped with an insulating material, and are then arranged in lengths of wrought-iron tube, which are afterwards filled with an insulating compound to prevent all contact between the wires and the iron tubes.

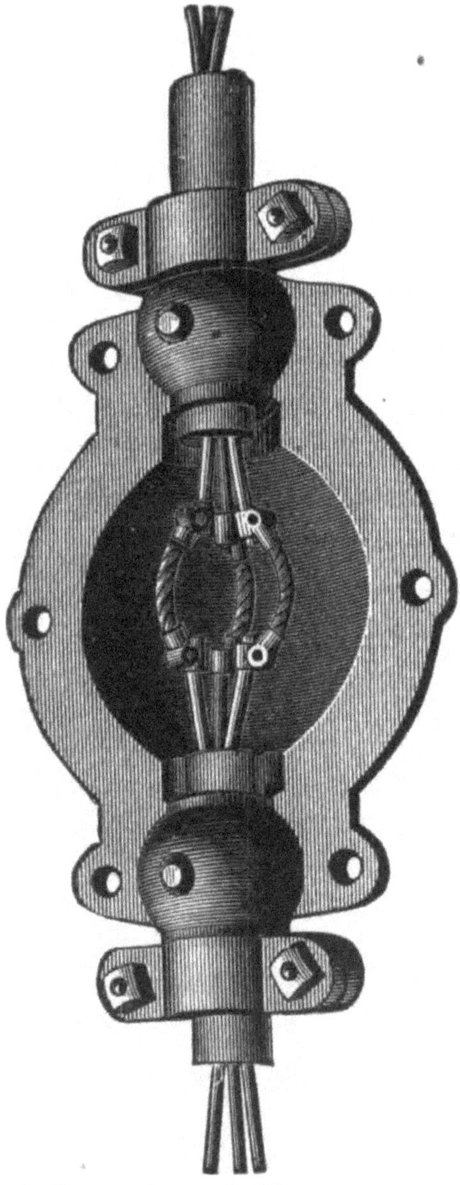

FIG. 15.—Form of Junction Box for Coupling Lengths of Tubing.

These tubes when laid are connected together by means of cast-iron junction-boxes, into which their ends are secured,

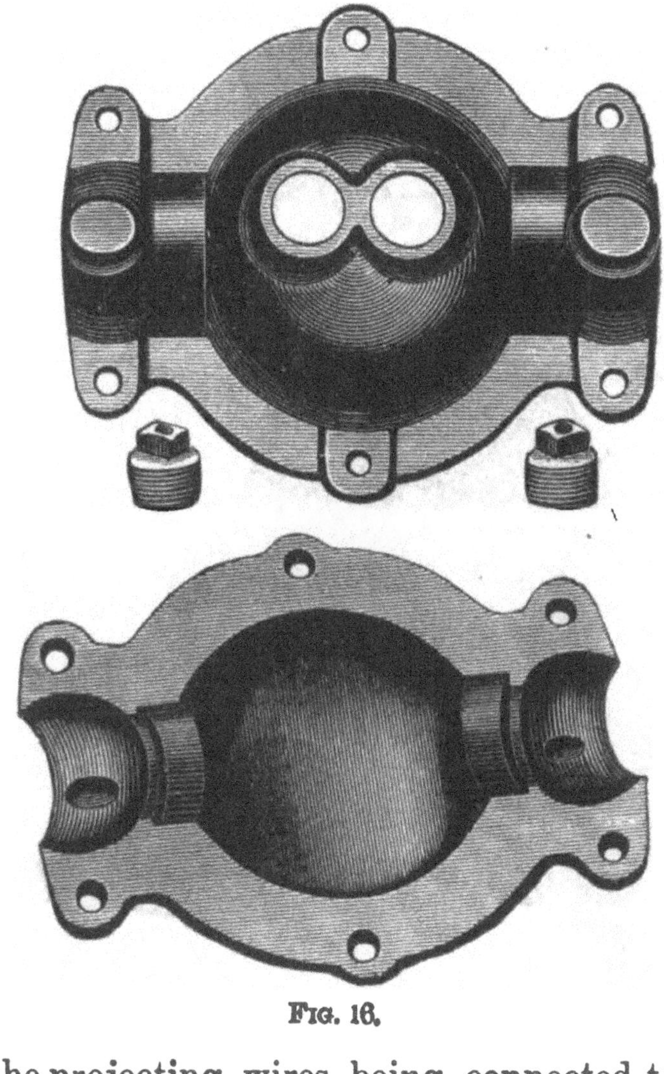

Fig. 16.

the projecting wires being connected to each other by flexible connections. These boxes are made in halves, the upper half

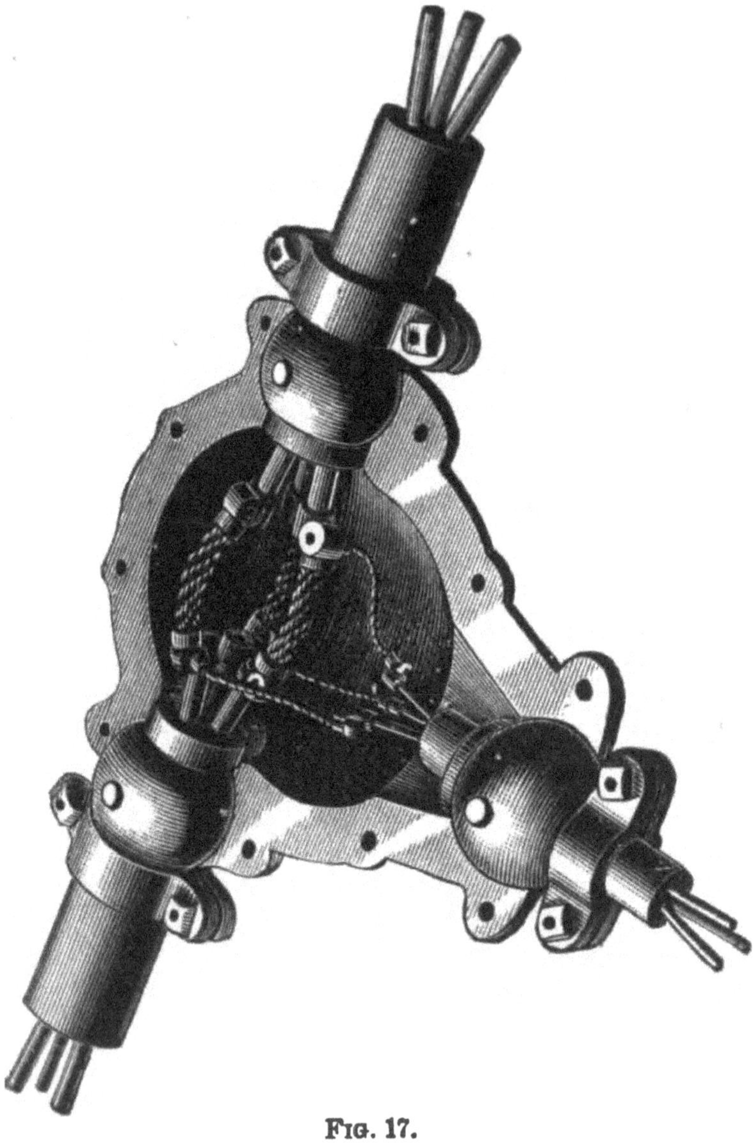

FIG. 17.

being provided with holes which are

finally closed with screw plugs after the halves have been firmly secured together and the interior filled with insulating compound.

In Fig. 15 we have a view of an open

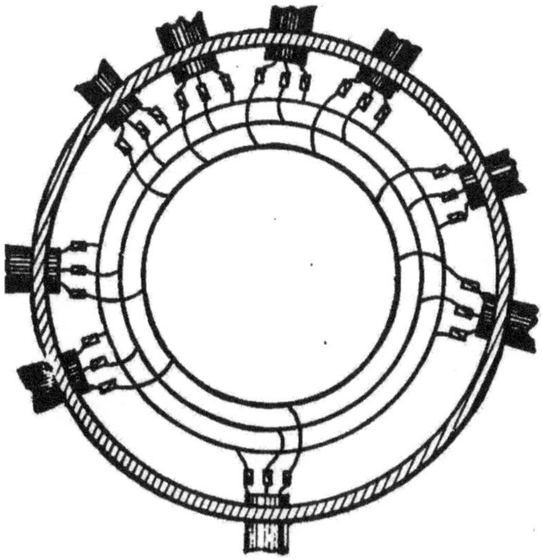

FIG. 18—This diagram represents a man-hole, and shows the interior connections and distribution between feeders and mains.

box showing the method of joining the wires together with the flexible connections; and in Fig. 16 is shown the upper and lower halves of the box with holes and screw plugs as described above. Fig. 17 shows a branch box having small wires leading therefrom. These boxes are used where wires are to be led into a building, or to supply a street branch. Fig. 18 shows the method of running several tubes into a man-hole and connecting all together there.

To return to the station, Fig. 19 shows the general arrangement of the electric devices therein. Here are shown three units or couples of dynamos; each individual dynamo having a regulator for controlling the amount of current passing through its magnet coils.

The three heavy lines running nearly the full length of the figure are what are termed the "Bus" wires, and are marked respectively, positive, negative and neutral, as shown by the signs attached to each.

The object of these wires is to receive the total current delivered by the units, and conduct the same to the feeders for distribution to the outside mains. It will be observed that one wire from each of the couples of dynamos passes to the neutral bus, while the other two wires pass to the positive and negative buses, re-

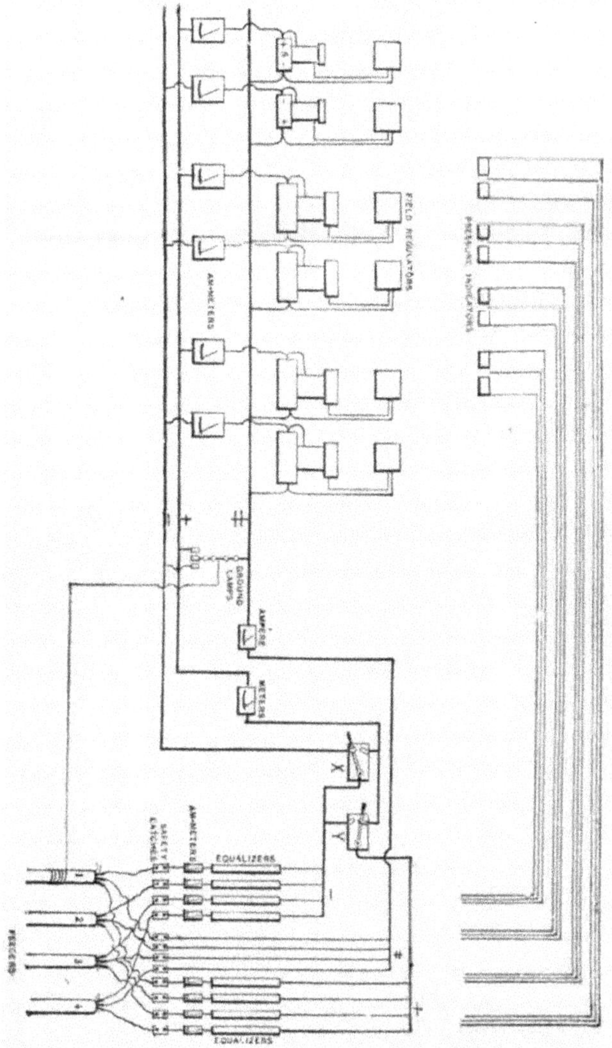

Fig. 19.

(*To face page 52.*)

spectively, through "Am-meters" which show the amount of current being delivered by each couple, while two other ammeters are arranged in the bus line to indicate the total amount of current flowing out to the lamps. Beyond the ammeters are placed what are known as "Changing Switches," by means of which either the positive or negative circuit may be broken at will, so as to make both outside wires of the system either positive or negative; the neutral wire serving in such a case, either as positive or negative according to the way the switches are thrown.

This arrangement of the wires is rarely brought about except in case of an accident disabling some of the engines or generators when the load on the wires is very light.

To their proper bus are connected the feeder wires enclosed in tubes, 1, 2, 3, 4, leading to the junction boxes similiarly numbered. These feeder wires are connected with the equalizers before described; the neutral wires, however, being without them as, when the circuit is properly balanced, no current passes over these wires. In addition to the equalizers the positive and negative feeders are supplied with am-meters so as to guide the attendant in regulating the current in each, and all have "Safety-catches" composed of strips of a metal melting at a very low temperature.

Any excess of current passing over the line of feeders, either to or from the station, will melt these safety-catches before any damage can be done either to the lamps outside or to the apparatus within the station.

The number of lights in operation upon a circuit varies at different hours, and this variation must be provided for; that is, only enough current should be generated to supply the lights in actual operation.

It would be far from economical to run a station up to its full capacity throughout the time when lamps are burning, as different classes of buildings require light in different measures, and while some use light only an hour or two, others require it for several hours.

These conditions are met by the attendant at the station who, by means of the various devices for regulating the current, keeps a pressure upon and a quantity flowing over the wires only sufficient for the number of lights in use at any given time.

As the lights are turned on or off the current increases or diminishes, and up to a certain limit this change is met by using the regulators.

When the current becomes too great to be controlled by these, one or more units are thrown out of operation; that is to say, one couple of dynamos is stopped, and then another, and so on as the lights are gradually turned off. On the other hand, as lights are turned on and the supply of current increases, one couple after another is put in operation to keep pace with the demand.

To make this clearer, we will say that two dynamos are started late in the afternoon a little before the time when lighting generally commences. If no lights are turned on there will be no current generated, because we shall have what is termed an open circuit; that is, there will be no connection between the outgoing and incoming, or what is known as the positive and negative wires, as there is when lamps are in circuit.

As soon, however, as a lamp is turned on, a bridge is formed connecting the main wires and permitting the current to flow over the completed circuit, which includes the generator, the latter becoming active the moment the circuit is closed by the introduction of a lamp. A glance at the volt meter shows us that we have a pressure greater or less than 100, which is the pressure we require; and by operating the resistance boxes connected with the fields of the dynamos, we secure the proper pressure, and the current being proportional to the pressure divided by the resistance of the lamps in circuit will take care of itself.

As more lamps are turned on our pressure begins to fall below 100 and, we have to manipulate our field-resistance boxes so as to permit more current to flow around the field-magnets, thus increasing our current and raising our pressure to the proper point.

This is continued until the number of lamps in circuit nearly equal the capacity of the pair of dynamos in operation, when two others are started, and are permitted to get up to the proper speed before they are connected with the general circuit; for the pressure of the current from any dynamo varies with its speed; and should a dynamo be connected to the general circuit before it had attained the necessary speed to give its current the proper pressure, the current from the dynamos already in operation would pass through it and cause its armature to rotate without producing any current; on the contrary, it wonld absorb an amount of current proportional to the power necessary to turn its armature, and whatever that armature might be connected with by belting,—this condition is known as making a motor of a dynamo.

Although the dynamos about to be turned on may be running at full speed, and ready as soon as thrown into circuit, by what are known as switches, to generate a current of the proper pressure, they need not necessarily be prepared to generate a large amount of current, for the reason that the field-resistance boxes may be so arranged as to allow only a very small amount of the current generated to pass around the fields; so the couple, when they are finally thrown into circuit, add but little to the amount of current going forth over the line but, as soon as they are fairly working, the attendant by means of the field-resistance boxes of the four dynamos regulates them so as to have each generating its proper portion of the current or, as it is generally expressed, carrying its portion of the load. As the load—that is the number of lights—increases, these four dynamos are brought up as near their full capacity as is desired, and then others are thrown in, and so on, until the station is working at its full capacity; the regulators meanwhile being operated to keep the sides of the system in balance.

The operation of decreasing the output of the station is somewhat similar to the foregoing. The excess of current, as lights are turned off, being taken care of by introducing the resistance of the regulators up to the proper point, and then operating upon the fields of the dynamos, until the current necessary for the lamps in circuit falls to a point where it is expedient to cut out a pair of dynamos.

The couple to be cut out of circuit, have their field-resistance boxes manipulated until they are generating only enough current to keep their pressure up to the proper point to prevent the main current from turning them into motors, when they are thrown out of circuit and the engine driving them is stopped; the other dynamos having meanwhile been so regulated as to divide the total load between them, and this operation is repeated as day draws on, until the entire station is at rest.

That the operations just described should be intelligently and carefully performed, is of the greatest importance, as each lamp when burning represents a given amount of coal being consumed at the station; and as the light is sold at a predetermined and stated price, just as gas is, it is necessary, in order to have the electric light company's books balance in favor of the proper parties, that no coal should be burned unless the light is in actual use. This can be accomplished only by reducing the outgoing current as rapidly as it can safely be done when the lights are turned off, and increasing it only as fast as is absolutely necessary when lights are being turned on; for, the greater the current going forth, the more dynamos in operation, the larger the number of engines required to run them, the more steam at a given pressure must be furnished to the engines, and the more coal of necessity burned to keep up the supply of steam.

From this you will see that there is a close relationship between the incandescent light and your grate fire; the former being the energy of the latter in a form almost identically the same, in both instances we have incandescent carbon, the ultimate object in one case being heat, and in the other light.

Having proceeded thus far, let us glance back over the road by which we have come and see that we fully comprehend the relation, each to the other, of the several devices necessary for the creation and utilization of the electric current for the purpose of lighting by incandescence.

Strictly speaking we should start at the coal pile; but were we to whittle our stick to so fine a point we might be tempted to go back beyond the glacial period, when our coal fields existed as the superabundant vegetation of a tropical clime. It will, however, suffice us to confine ourselves to this age of Light;

and lest we be tempted to turn aside, into the by-paths of speculation and geological research, we will start with the furnace, that indispensible adjunct of the steam boiler.

Having generated sufficient steam for our purpose we open the valves and let it enter the cylinder of the engine, where its energy is transmuted into the rotations of the fly-wheel and these rotations are in turn transmitted—through the belt connecting the fly-wheel to the pulley on the armature shaft,—to the armature of the generator. Here our current is generated and hence it goes forth over the feeders to the thousands of lamps scattered over the district being lighted.

The glowing carbon in our furnace, has compelled by its heat the water in our boilers to assume the form of steam; the pressure of which in our engine has developed motion; and this motion, transmitted to our dynamo, has there taken the form of electricity, and flowing forth over the line, has in the lamps by its energy produced again both heat and light.

We know, in a general way, the use of each device involved in this system of electric lighting; it may, however, be interesting to know something more definite as to the form and arrangement of

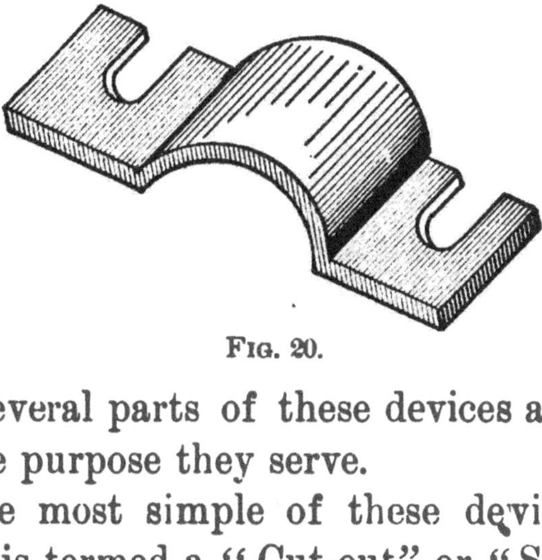

FIG. 20.

the several parts of these devices as well as the purpose they serve.

The most simple of these devices is what is termed a "Cut-out" or "Safety-catch." This is made in a great variety of forms according to the position in which it is to be used and the quantity of current it is to carry. That shown at Fig. 20 is intended to be attached to, and connect the metal portions of a feeder or other switch for carrying heavy currents; it is composed of two end-pieces of copper connected together by a strip of alloy composed of a number of metals so combined as to cause it to be readily acted upon by a very low degree of heat; that is, to be melted by a current far below what would have an injurious effect upon the wires or other portions of the electrical apparatus in circuit with it. The copper end-pieces are provided with slots for the purpose of passing about the shanks and under the heads of the screws by which they are secured to the switch. A space is left between the parts of the switch to which it is secured, so that the safety-catch forms a bridge from one portion of the switch to the other. Should the current become too great for its carrying capacity, the alloy would melt and fall down upon the non-conducting base upon which the several portions of the switch are mounted, thus breaking the circuit and cutting off the current until suitable provision be made for its control, when another safety-catch is substituted for the melted one, and all is ready for continued operation. When wires are led into houses, a different form of safety-catch is used. That shown in Fig. 21 is intended to be placed in the line immediately within the building before the wires pass through the meter. It is arranged for three wires, and is formed of wood, porcelain, or

FIG. 21.

other suitable insulating material; in it are arranged three sockets, similar to those in which lamps are placed, composed of a screw shell, at the bottom of which is fixed a metal button, the shell and button being separated from each other by the insulating material of the block in which they are mounted. On either side of these sockets is a screw, one of which is connected with the button, and the other with the shell.

When this cut-out is in place, the three wires entering the building are secured to it by means of the screws on one side of the sockets, while the three wires leading to and through the metre are secured to the screws on the opposite side of the sockets, the circuit remaining open however until a plug is screwed into each of the sockets. This

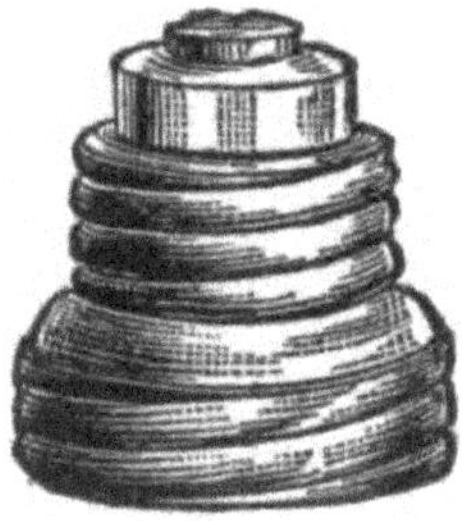

FIG. 22.

plug, shown in Fig. 22, is formed of glass, and is hollow. On its bottom end is fixed a metal button, which is connected to a screw shell arranged on the outside by a strip of the fusible alloy before mentioned, located in the hollow space, in the interior of the plug; the top being provided with a metal screw cap perforated to allow the gases to escape, if for any reason, the strip of alloy is melted by an excess of current.

When the plugs are screwed into the sockets in the cut-out, their metal button makes contact with the metal button of the socket, while the screw shell arranged on their outside fits closely into the screw shell of the socket, thus allowing the line-wires connected to the cut-

FIG. 23.

out to communicate with each other through the fusible piece contained inside of the plug. Farther on in the house where wires branch off from the main circuit into the several rooms, cut-outs like that shown in Fig. 23 are used, the three wires of the main circuit passing along them in the slots between the sockets, secured therein and connected with the sockets by the screws shown in the cut, while the branch wires are led off from the screws shown between the sockets on either side of the cut-out. Similar cut-outs, with three sockets on each side, are provided for three wire branches.

In the two wire cut-outs the buttons of two of the sockets are connected to the central or neutral wire, while the buttons of the two other sockets are connected with the positive and negative wires respectively, thus disposing the lamps on the branches equally on both sides of the system. In the three wire cut-outs, two of the sockets connect with each of the three main wires.

These cut-outs are called two-branch cut-outs as they admit of a branch circuit being led off from both sides of the main circuit. Cut-outs similar to these are provided to meet all the requirements of distribution for interior lighting.

Switches, like cut-outs, vary in form according to the purpose for which they are to be used. Those with which the public are most familiar are shown in Fig. 24.

They consist of a spindle mounted on a non-conducting base, and having arranged upon it a metallic piece, which, by turning the handle attached to the spindle, is brought into contact with metallic pieces to which the line wires

FIG. 24.

are secured by suitable binding screws thus closing the circuit.

The spindle is turned against the stress of a spring, so as to break the circuit quickly, when, by rotating the spindle, the bar mentioned is released from the metallic line wire contact-pieces. These switches are made of different sizes for carrying different strengths of current.

HOUSE LIGHTING.

In laying the mains along a street, suitable junction boxes for branch circuits are inserted in the line at points opposite the buildings in which lights are to be used, whence wires of a proper size to carry the current to be used are led into the buildings. Taking one of these branch circuits, or house "Services," for example, we lead it underground through a tube similar to those laid along the street until the wires enter the cellar or basement of the building. Here they are connected to a three-wire cut-out, and are then led to the meter. This cut-out is to provide against a larger amount of current than is necessary for the number of lamps to be used entering the house. The meter is also proportioned in size to the number of lamps employed. After passing through the meter, the wires are led to an elevator shaft, hallway, or suitable space provided in the walls for this purpose, as nearly central as may be, and are carried up to the topmost floor to be lighted. On each floor the wires pass through what is known as a pocket, which is generally a boxed-up space provided with a door for easy access to the enclosure.

In this pocket are arranged a number of cut-outs and switches corresponding to the number of branch circuits to be run to the several rooms on that floor. At this pocket, by turning the proper switch, lights may be turned on in any room before entering it, or the current may simply be turned on, and the lights on that branch be cut in one by one by turning the keys with which each socket is provided.

For large chandeliers, a switch is generally arranged against the walls, at some suitable point in the room by which all the lamps on the chandelier may be turned on at one and the same time.

Arrangements are often provided by

which lamps may be turned on in the hall at the front door, and extinguished from the top of the stairs, thus allowing one to have a light before them in whatever part of the house they may choose to go. This is a matter of economy as well as convenience, for it allows of lamps being lit only when and where needed, enabling occupants to use light only in the rooms occupied, while the rest of the house remains in darkness that may be dissipated in an instant at the will of any individual occupant.

Lights outside as well as in are equally under control, and the coming or retiring guest may be lighted to the farthest point of the most extensive grounds without the necessity of the host being in any way exposed to the weather. To crown all, these luxurious conveniences are furnished at a price to place them within the reach of persons of but ordinary means. After the plant is once installed there is only the cost of the light actually used. No leaking taps, or defective joints, no smoke or smell, or dirt—

those unpleasant features inherent in all other forms of artificial illumination.

The various devices once placed are likely to last under ordinary usage as long as the householder will probably live to enjoy them ; the only exception to this general statement being the lamp itself which, with the proper current for which it is constructed, is guaranteed to last one thousand hours. As the lighting company replace all lamps burned out and, as they control the current to which they are subjected, there is no reason to doubt that their life will be prolonged to the utmost limit consistent with their being burned at their full "Candle-power," which is that of an ordinary gas jet, from a two-foot burner.

The term candle-power refers to the standard wax candle, which is the unit of illumination adopted in grading electric lights ; the lamps commonly used in dwellings being rated as giving a light equivalent to that of sixteen standard candles.

In addition to the large number of advantages possessed by the electric light over all other forms of artificial illumination, it may be said to be far in advance of all others in the readiness with which it may be adapted to all forms of decoration.

It may be mounted in chandeliers, supported in brackets, fixed directly to walls and ceilings in any and every conceivable position, or allowed to hang at the ends of flexible cords. The globes may be of all colors, and the best talent of the civilized world has been levied upon the production of beautifully designed fixtures to receive them, as well as globes, shades, and reflectors to modify the lights and enhance the beauties of its effects. Like the light of the sun, it beautifies all things on which it shines, and is no less welcome in the palace than in the humblest home.

We have endeavored in the preceding pages to state the fact clearly that in the Edison three-wire system of incandescent electric lighting the pressure upon the wires is kept constant. A constant pressure implies a constant current, to measure which the Edison meter has been invented. Mr. Edison is the fortunate possessor of that rare combination of faculties, superior inventive ability, supplemented by unusual business qualifications and, to his intimates, the fact that he has invented a device is a sufficient guarantee of its commercial utility.

The foregoing statement holds good with regard to the Edison meter for the measurement of the electric current supplied to one or more incandescent lamps, and is called up by the fact that upon the introduction of this meter the claim that it was practical in its operations and sufficiently accurate to form the basis for the charges to be made the users of light, was ridiculed by many who would be well pleased had they withheld their opinions till time and use had paved the way for a more just decision upon the merits of the piece of apparatus in question. Time and daily use have proven the Edison meter not only reliable but beyond that, a standard for charges satisfactory both to the furnishers and users of the electric light.

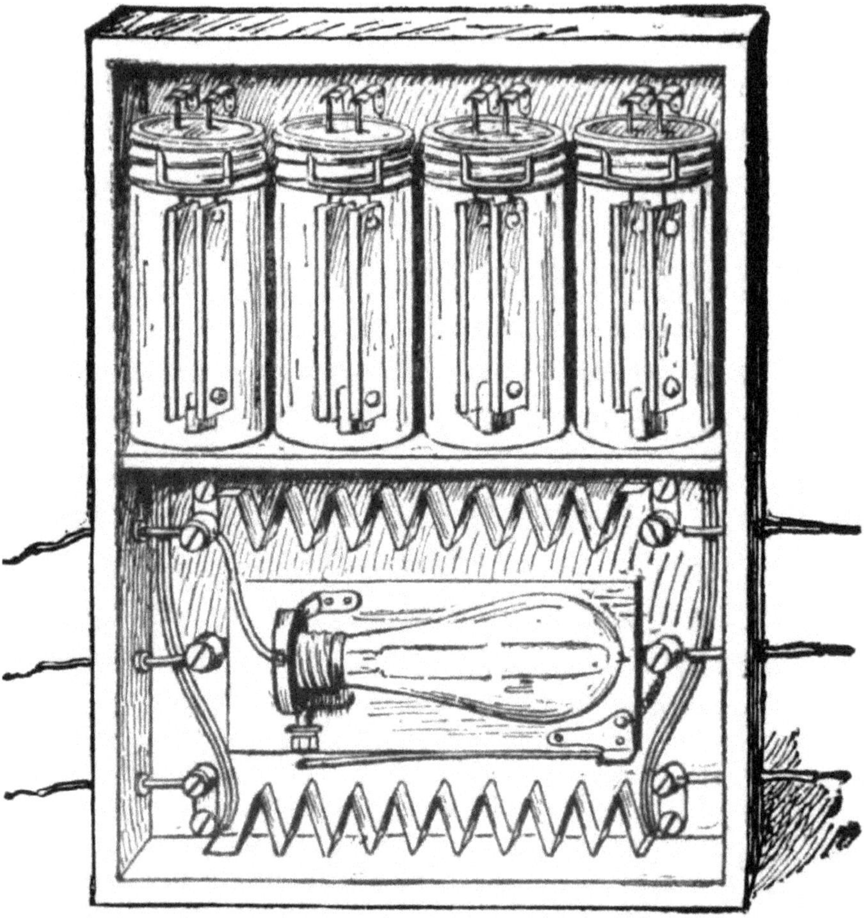

FIG. 25.—Three-wire meter, open.

Its operations are based upon the well-known electrolytic action, which causes two zinc plates immersed in a solution of zinc sulphate to vary their respective weights in proportion to the current passing through them. This will be understood by referring to the cut Fig. 25, which shows a three-wire meter with the front of the case removed.

The cut shows the meter case divided by a partition arranged across its centre, the space above it being occupied by the bottles or jars, containing the zinc plates, immersed in the zinc sulphate solution.

Fig. 26.—Meter bottle.

These bottles, Fig. 26, are provided with glass tops, held in place by screw ring, the glass tops having holes in them, through which pass the copper stems of the zinc plates, to engage with the spring clips arranged above the bottles, and serving both to hold the bottles in place and to connect the plates with small wires, leading to the space below the partition. The zinc plates are kept a proper distance apart by hard rubber fittings, which are also arranged to keep the plates from resting upon the bottom of the bottle.

The small wires before mentioned as passing from the clips above the bottles to the lower space of the meter box lead the current to the zinc plates in such a manner that it must pass from one plate to the other through the zinc solution. In doing this the current detaches particles of zinc from one of the plates and transfers them to the other, thus decreasing the weight of the first plate. One ampère, acting for one hour, will deposit a known amount of zinc upon the negative plate in the bottle; and it is by removing and weighing the plates that the amount of current which has passed through the meter is ascertained.

The bottles are arranged in the meters in pairs, two pairs being used in the large sizes of three-wire meter to insure by comparison greater accuracy in determining the amount of current which has passed.

Through the lower space in the meter pass the conductors leading into the house, or rather they pass into the meter box and are secured in binding posts located on each side just within the meter. These binding posts are connected together, by what are termed "Shunt's," in this case, composed of broad strips of German silver, carefully graduated as to their resistance, so the $\frac{1224}{1225}$ of the current passing over the wires will pass through them, while $\frac{1}{1225}$ of the current will pass through each of the bottles, by means of small wires which may be seen passing up on one side of the meter and down on the other, connecting at both ends to the binding posts of the lower conductor. The upper and lower conductors alone are provided with shunts, the middle or neutral wire having none but passing directly through the meter behind the "Thermostat," the lamp upon which, is arranged between the neutral and upper conductor. The purpose of the thermostat is to prevent the freezing of the zinc solution in the bottles, which it does by the automatic lighting of the lamp upon it; the heat from this lamp being sufficient, in the confined space within the meter, to accomplish the purpose desired. The thermostat, Fig. 27, consists of a block of wood to be secured against the back of the meter. Attached to one end

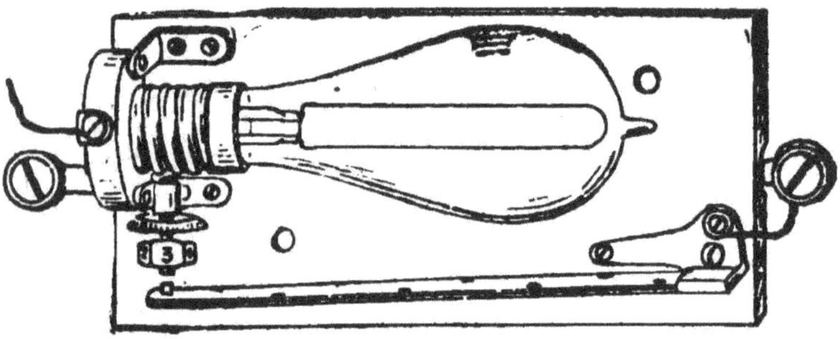

FIG. 27.—Thermostat.

of this is a lamp socket, one wire from which is attached directly to the conductor leading through the meter, the other being connected to an adjustable metallic screw on the side of the socket. Near this screw is one end of a strip of brass to which is riveted a companion strip of steel, the opposite end being secured to the main block fixed to the back of the meter. From this latter end passes a wire, which is secured in a binding post located at one end of a piece of metal passing behind the thermostat, to connect the two ends of the neutral wire entering the meter box at opposite sides, together. A lamp having been screwed into the socket, it is ready for operation.

The end of the brass-steel strip, opposite the screw on the lamp socket, is provided with a contact-point which will rest against the end of the screw under proper conditions; that is, when the temperature in the meter falls below a certain point, the brass on the brass-steel strip, contracting more than the steel, will cause the strip to curve, thus bringing the contact piece on its end against the lamp base screw, and, completing the circuit through the lamp, cause it to glow and disseminate the necessary heat.

It is a quality of sulphate of zinc to increase its resistance as it cools, and consequently the resistance of the bottles is ever varying with the change in the temperature about them. As an offset to this seeming defect in the apparatus small spools of copper wire, accurately measured, are placed in the circuits of fine wire leading to the bottles. The conditions which increase the resistance of the bottles, decrease the resistance of the spools; and thus one offsets the other to maintain a constant resistance.

These meters are in use both in this country and abroad in large numbers, and give satisfaction wherever used. They are simple and inexpensive in construction as well as economical and accurate in operation ; qualities which have directly influenced their adoption in all the large central stations throughout the country.

DESIGN AND OPERATION

OF

INCANDESCENT STATIONS.

BY

C. J. FIELD.

DESIGN AND OPERATION

OF

INCANDESCENT STATIONS.

By C. J. Field.

I DESIRE to present to you a brief review of the present and prospective fuure of central power plants in the larger cities, taking as an illustration one of the more recent types, describing its general arrangement, then proceeding to the consideration of its initial cost, earning capacity, output, operating expenses and economy, and, in conclusion, trying to indicate the immediate future development in this class of work.

CENTRAL STATIONS.

The immediate points to be considered and carefully weighed in the designing of central power plant for a large city are many, and they should receive careful survey before any work is proceeded with. We will briefly summarize them as follows:

First—Recognition of the importance of safety and stability in operation.

Second—Obtaining the true economy of output under all conditions.

Third—Installing of plant in a building entirely suited to the working of the same, and as far as human ingenuity can provide, proof against destruction.

Fourth—Adaptability to proper and economical working of the plant.

Fifth—Division of the generating power into the proper number of units for the safe and reliable operation of the plant.

Sixth—Flexibility of system; that is, adaptation to furnishing current for light, power and other sources of revenue, the obtaining of the largest return per dollar invested, and not carrying to excess for the mere sake of engineering in any part of the plant but the obtaining of proper results therefrom.

Seventh.—Not installing the plant for mushroom growth, but laying it out for comprehensive business, thereby securing at as early a date as possible the entire confidence of the invested capital.

A true and careful consideration of these points will prevent trouble later on. Much of the trouble of stations at the present time in their standing with the community is due to neglect of this point, and the majority of their failures as well. We have got to recognize the fact that the public, to a certain extent, have become prejudiced, in a measure, somewhat unjustly; but this is all the more reason for better and more conservative management and giving them good construction. No more inviting field is offered for either investing capital or good engineering than a central station for lighting, power and railway work.

A REPRESENTATIVE STATION.

I propose to take as a representative type, showing the present development and first-class work, the station of the Edison Electric Illuminating Company, of Brooklyn, which was completed last fall, and is now in successful operation.

In the arrangement of this plant there was somewhat of a departure from previous general practice in this line, the company trying to secure the benefit of past experience in the larger stations of this class, both in the arrangement and kind of apparatus used, trying to secure at as economical a cost as possible the best plant for the purpose.

The boilers and engines are located on the first floor, the engines being on the front half, and the boilers at the rear, thereby bringing everything in this part directly under the eye of the chief engineer, making it much better than where the boilers are located two or three stories up; this was obtained by spreading out a little more on the ground. The boilers are Babcock & Wilcox's largest type of sectional water-tube boilers. The engines are 300 h. p., compound, horizontal, automatic engines, manufactured by the Ball Engine Company. Each engine is directly belted to two generators.

Ascending to the second floor, we reach

the electrical part of the plant. Here are located in the front part of the building, directly over the engine-room, twenty-four dynamos, each with a capacity of 750 ampères and 140 volts. Each dynamo weighs about eight tons. Overhead travelling cranes are installed here and in the engine-room for ready and quick handling of all apparatus. Through the centre of the dynamo-room is located the electrical gallery. From here are controlled the workings of all the dynamos and other apparatus, also all outside lines. Everything in connection with handling, generation, and furnishing of current is directly under the eye of one man in this gallery, and from which he has a general view of the dynamo-room floor and the workings of the dynamo, a second man being on the floor to see to the bearings and brushes. From this gallery run all the feeders, which connect into the network of mains, covering over an area of about one and one-half miles square. The ampère meters are located on each feeder, so as to show the load in each part of the district. This plant maintains its distribution and regulation thereof by balancing within itself. No feeder equalizers are here used for feeder regulation; the uniting and tying up of the system, together with the use of the auxiliary bus effects this regulation. All circuits of this plant are underground, there being about twenty-five miles of underground conductors. These have given perfect satisfaction and reliability in their workings, maintaining to-day an insulation on the system as a whole of over half a megohm.

On the rear of the second floor are located the coal storage, water-tanks and feed-water heater. On the top floor we have the offices, supply rooms and workshops of the company. Returning downstairs again we find in the basement ash-pits, smoke flues, pump-rooms, two large coal-storage vaults, giving a total capacity for storage of over 1,000 tons, air-blast for forced draft and other details in connection with the steam plant.

We have, therefore, here, in a building

75 x 100 feet, apparatus and all departments complete for the generation and supply of current and power for a capacity of 40,000 lights, or the equivalent in light and power, and so arranged as to secure, as far as can be foreseen, continuous working of the plant and entire reliability in the furnishing of its current. It is only thus that we can hope to obtain a business and establish ourselves on the commercial basis which gas light companies have placed themselves in the past years, and thereby secure to our stockholders the returns for which they have invested their capital.

Having thus generally outlined this plant, we will now turn our attention to the consideration of other points in connection with it. One of the most important items is the cost of such a plant. I give you below, in round figures, the cost as shown by the construction accounts and estimates:

Station building, complete, including all fittings, foundations, stacks, furniture, etc........................	$100,000
Real Estate..........................	36,000
Steam plant, including engines, boilers, pumps, heaters, piping, belts, etc....	50,000
Electrical plant, including dynamos and all electrical apparatus, as switches, etc...........................	40,000
Undergronnd system material........	115,000
Excavation and labor installing same..	35,000
General, including lamps, meters, tools, instruments, engineering and architectural expenses, wiring, services and office furniture...............	50,000
Total.	$426,000

This includes the entire cost for the plant as it stands to-day, which, as far as the building is concerned, is complete for the entire capacity. At present there is installed generating capacity of boilers, engines and dynamos, for one third ($\frac{1}{3}$) of the final output of the plant. The electrical apparatus is complete for the entire output, with a very few additions in the way of a few switches, etc. The underground lines have a capacity for 20,000 lights. The work necessary to complete the plant for its entire capacity would amount to about $200,000 additional. For this amount there has been obtained here a plant, which is considered equal, if not superior, to any of this class, and at a cost of twenty to thirty per cent. less than is expend

I will take up the next consideration of the operating expenses of such a plant. In order to place the company on an earning basis we have to secure to start with a certain number of lights or an equivalent in lights and power to clear the necessary general and operating expenses, which will exist regardless of the smallness of the load; in other words, we must have for such a capacity-plant not less than 5,000 lights with an average income of $8 per light per year to clear the general incidental and operating expenses. This figure we may consider as our unit of operating capacity. From this we can figure the increased earnings and profits for the larger number of lights connected. There exists practically a constant ratio of variable and fixed operating expenses. By variable expenses we mean those obtained on a variation in load and increase of business. This includes coal, oil, lamp renewals, and small increase from time to time in the amount of labor employed. The fixed expenses include those items which remain practically constant under varying conditions of income. A careful analysis of all the items covered in these expenses in such a station as this one, gives the following result: That the fixed expenses are seventy-five per cent. of the whole, and the variable twenty-five per cent. approximately; or, in language which may appeal more directly to you, if we double our income or business, we only increase our expenses 25 per cent. This shows that a station's possibilities and profit lie in increasing this business from the unit point.

The average income per light in stations of this class varies in different parts of the country and with different loads. We have obtained a load diagram, taken from this station, which gives a fair idea of the changes and variations here taking place. The maximum number of lights lighted at any time in proportion to the number connected is very good for such work and shows a good class of business. The load diagram through the day, however, shows a new station on a clear day with a small number of lights lighted, and power work only just commencing. It is this power work that wants attention, and the securing of which means the bringing up of this average load during the daytime to a good paying basis. The curve for the evening hour indicates a good, broad, solid load, which shows the combination of six o'clock business with the addition of a good solid evening load. Many stations after reaching the maximum point around six o'clock, rapidly fall off and never regain that point again for the evening. Then we have the illustration of clubs, theatres, churches, concerts and residences, lighting after supper, bringing up the load to its maximum point between eight and nine o'clock. The average load for the twenty-four hours, which is about twenty-five per cent. of the maximum is a very fair one, and with the addition of the day load that will come on with the addition of power, it makes a model load diagram. The general run of stations shows an average for the twenty-four hours of from twenty to forty per cent. of the maximum load; the latter figure is very seldom reached. The writer knows of one station in which we have the latter figure, and where, if we eliminate one short half-hour around six o'clock, we have the remarkable showing of seventy-five per cent. average load for the twenty-four hours.

We have taken some remarkably fine indicator cards, which show the workings of the engines of the steam plant in this station. The division of work shown on the cylinders is as follows:

I. H. P. head end high pressure cylinder.....................63 8-10 h. p.
I. H. P. crank end high pressure cylinder.....................60 2-10 h. p.
I. H. P. crank end low pressure cylinder.....................59 2-10 h. p.
I. H. P. head end low pressure cylinder.....................55 2-10 h. p.

Making a total I. P. H...238 4-10 h. p.

with a boiler pressure of 110 pounds, revolutions 223 and load 1,200 ampères, which is the equivalent of 2,725 lamps; therefore giving for indicated power furnished eleven and one-half lamps per horse power. This is for power developed, making no allowance for friction of engine and dynamo. The friction of the engine is less than five per cent. of its normal capacity.

Having in the above given a general outline of this plant, its cost and operating expenses, I now wish to call your attention to the points in connection with the type of engines, boilers, dynamos, underground system, etc., to be adopted in a station of this class.

CONCLUSIONS.

We will first consider the question of the engine. As already stated, in proposing the engine power of a station of this kind, we first have to consider the question of using either the Corliss or high-speed engine. Regarding the use of Corliss engines in a plant of this kind, we are frank to state our objections. Excessive first cost, ponderous machinery, counter-shafting, pulleys, clutches, etc., lead us to believe that these things are unnecessary when the problem is carefully considered from an unbiased standpoint. What we are after is results; not theory, but actual practice. Assume, for the sake of argument, that we can save five or ten per cent. in steam economy; if this is obtained at a cost, the interest of which amounts to more than this, we are obtaining it for no good whatever; furthermore, there are many other problems in electric light stations which we have to carefully consider in this question of steam plant, one of which has been enumerated before, viz., the question of reliability in operation, and always being ready for service. One of the latest types of stations combining arc and incandescent, where we have the Corliss engine in all its perfection of detail and apparatus, is that of the Narragansett Company, in Providence. If one will carefully look over this plant, as I had the pleasure of doing a short time ago with others, and consider all these problems carefully, and then examine a station similar to the Brooklyn one, I think he will be forced to admit this fact. We want to obtain our power as direct from the engine to the dynamo as possible, and at the same time as cheaply, and obtain the best economy under the variable loads we are going to have. We cannot design our plant for that capacity which is reached as shown on our load diagram, for only a short time in the twenty-four hours, but we must so design it to give this result for the average that we have during the twenty-four hours. Even where we have a more constant load, as in exclusive arc lighting on municipal circuits, I think even here we need to carefully consider the problem as well.

High-speed engines, so called, although they are not in piston speed any higher than the Corliss, but merely in rotative speed, have shown a considerable development and marked advance in the past year, and the next year is going to see even more development in this line. Owing to better workmanship, better designing and building than formerly, the prejudice which largely existed among old engineers against this type of engine is rapidly wearing away. With the single cylinder engine under variable load, we often obtain poor economy, but, as compared with the Corliss, under similar conditions, allowing for the discrepancy in price, the result is not so disparaging. Now they are going further, and building compound and even triple expansion engines of this class. In the Brooklyn station we have the Ball, one of the representative types of this class of engines, being horizontal, compound engines.

These engines are built for high economy and economical work, and the guarantees made on them I think compare favorably with the guarantee on compound Corliss engines; at least, three or four of the manufacturers of this class of engine stand ready to-day to guarantee from 22 to 25 pounds of water per indicated horse power per hour. I do not know of any Corliss manufacturers who are willing to do any better. This is for non-condensing; condensing from 17 to 18 pounds of water per horse power per hour. Engines of this class are as well built now in workmanship, and as reliable in operation as can possibly be desired. Added to this, we have the advantage of direct connection to our generators, avoiding all the intricacies of shafting, etc., and the unreliability they entail. Tests made at the Brooklyn station have shown that the engines have actually come up to the guarantee made on them, and that the plant there is showing, as compared with single cylinder engines, an economy of coal per unit of output of from 25 to 30 per cent. better. In a station of this kind, the actual coal consumed per unit of output at the dynamos is considerably larger than is shown in a direct test where we charge the engines only with the coal it uses directly. The weekly records from stations of this class charge the horse-power output with all the coal used by the engine, pumps, condensers, well pumps, cleaning fires, blowing-off boilers, etc., and where the former item is about three pounds of coal per horse power per hour, we have in the latter case, making no allowance for the engine running empty, a result fifty per cent. greater than this. Economy in this line, however, is not going to stop at compound engines, as there are being built by at least two manufacturers, triple expansion high-speed engines. (Mr. Field here showed views of such an engine, imported from France by Mr. Edison.)

Something similar to this is what we may obtain to-day, if encouragement is offered, from such engine manufacturers as Armington & Sons, Ball Engine Company, McIntosh & Seymour and others. We are coming to a recognition of the fact that if we want the high economy we can obtain it as cheaply and as well, not to say more cheaply and better, with an engine of this class as with engines similar to those installed in the Providence station. In guaranteed economy it will equal the Corliss engine, as installed in the Narragansett station, and give it to you under a wider range of load.

In connection with the triple expansion engine mentioned, we have to consider, again, the problem of the dynamos to be used. We can stay as at present, and belt our dynamos, but I believe that the next large incandescent station will not only include compound or triple expansion engines of 300 or 400 h. p., but will also have multi-polar dynamos, one or two being directly connected to the engine. By this I do not mean belted, but direct shaft connection through a flexible coupling. This of course, necessitates the multi-polar machine, in order to secure the output with a slower speed. Engines and dynamos of this type can be installed in the space at present occupied by the engines alone. This means not only economy in building and real estate, but also in operating expenses.

In regard to boilers for such a plant,

we do not know that we have any new economy to be hoped for in the near future. All we have to look for at present is improvement in detail of manufacture and the securing of better and dryer steam. We have two classes of boilers prominently before us for this work. We have, in general, first, horizontal tubular boilers, which we find in general factory use, to a large extent, throughout the country. Where we have plenty and cheap real estate, poorer attendance and moderate steam pressure, this class, in general, fills the bill. We find, however, that they are now even building them to work up as high as 125 pounds boiler pressure. When we come to construction of the boiler plant on expensive city property, where we are cramped for space, we are almost limited at once to some one of the types of sectional water-tube boilers. In the Brooklyn station we are practically limited to the consideration of this class, and we have not only 125 pounds but 150 pounds boiler pressure, and even higher. We have also the advantage of quick steaming under heavy changes in load.

We have to-day brought before us in the underground systems the consideration of what is to most of the companies their most serious problem, in the proper solution of which the best talent is being devoted. In the Edison underground system we have what is generally recognized as the most practical solution for circuits of less than 400 to 500 volts. We here obtain at a premium of cost the most flexible system and local distribution from house to house, which has no equal It enables you to take off services for local distribution from every twenty feet without in any way affecting the insulation on the main line, and being able at any time to disconnect these services and restore the main to its original condition. In any other system we have the problem of splicing and cutting of cables, which, at its best, is bad work. What we desire is not such a high insulation as good mechanical protection. As long as we can hold a moderate insulation with good mechanical protection, that is all we want. In Paris they have used bare copper conductors, supported on porcelain in a concrete conduit. This has worked satisfactorily in the main so far, but, of course, is very expensive. They are now proposing for all their increase, the Edison tubing for this class of work. In any system of cables drawn in we have the selection of a large class of conduits, but to my mind all we need and desire, as I have before stated, is mechanical protection for these cables, and the cheapest conduit that will afford this protection is all that is necessary. What is wanted especially is some system of local distribution for these higher tension circuits. The underground system installed in Brooklyn has a network of underground conductors in the mains and feeders of over twenty-five miles. This entire system is so arranged, distributed and connected in a network that, with a drop or resistance of 1 per cent. on the mains and 10 per cent. on the feeders, we are able to maintain in the system practically a perfect regulation in the distribution of the current. This system stands to-day representing one of the most complete and perfect examples of work of this class. The largest problems that we know of in underground work are the proposed new extension of the Edison Company in New York, and the underground system of feeders proposed to be installed for the West End Railway, of Boston. The copper alone for the latter amounts to over $1,000,000; and this question of the cost of the copper calls my attention to a fact which I desire to notice, that much of this question of bugbear on copper is uncalled for when we are considering the underground system. The entire copper used on the system in Brooklyn is less than one-fifth of the cost of the underground system first installed.

I do not desire to claim that the ideas for the class of work here represented and described hold or represent all the perfection to be obtained in central station work.

There are many points contrary to the ideas here outlined which are very desirable. I have merely tried to call your attention to what I consider good work in this particular line, and hope that it will result in bringing forth the discussion and additions which are very beneficial in the consideration of these problems in the results to be obtained, and I would only add a tribute to the powerful and master-mind whose work, from the commencement of this field of central station distribution, has covered the leading problems and points, and whose ideas to-day represent much of the good and very little of the bad problems which we have in this work. I refer to Thomas A. Edison, whose work commenced in this field on the old Pearl street station in New York, over eight years ago, when the majority doubted, and but few believed in its successful carrying out; while we find that station, until within the past few months, when it was partially destroyed, successfully working, and even antiquated as it was, earning large dividends. He has still continued actively to impregnate the work with his ideas from that day to this although he has not taken such an active part in its carrying out, but I think we may see him at no distant day again taking a hand in this work and bringing forth many new ideas in advancing the progress of the future.

THE MAXIMUM EFFICIENCY

OF

INCANDESCENT LAMPS.

BY

John W. Howell.

THE MAXIMUM EFFICIENCY

OF

INCANDESCENT LAMPS.*

By John W. Howell.

The word efficiency, when applied to an incandescent lamp, is used to designate the amount of energy required by the lamp for the production of a given amount of light; thus we say that a given lamp has an efficiency of three watts per candle, at sixteen candles, meaning, that to produce an illumination of sixteen candles we must supply the lamp with forty-eight watts.

The word efficiency, when applied to a prime mover or to any piece of apparatus that changes energy from one form to another, or which transmits or utilizes energy, has a well-defined meaning, and is used to represent the ratio of the energy of the useful effect produced by the ap-

* A paper read before the American Institute of Electrical Engineers, April 10th, 1888.

paratus, to the energy necessarily supplied to the apparatus to enable it to produce that effect.

An incandescent lamp transforms electrical energy into heat and light, so the use of the word efficiency to denote the watts per candle required by the lamp is not a proper one. To denote properly the efficiency of an incandescent lamp we must be able to separate the energy of the light produced by it from the energy of the heat produced. Then the ratio of the energy of the light to the electrical energy required by the lamp will be a correct expression for the efficiency of the lamp, and we will have to find some other word to designate the watts per candle. In this paper the word efficiency is used in its ordinary improper sense to denote the watts per candle required by a lamp when producing a given amount of light.

The efficiency of a lamp varies with its candle-power. The curve, Fig. 1, shows the rate of this variation for a particular lamp. At 5 candles this lamp has an efficiency of 6.7 watts per candle, at 10 candles it is 4.2 watts per candle, and at 20 candles it is 2.66 watts per candle.

Any statement regarding the efficiency of a lamp must therefore, be accompanied by a statement of the candle-power at which it has the stated efficiency; without this it is meaningless. There is nothing in an incandescent lamp itself that fixes its proper efficiency or in any way indicates what it is. The lamp from which the curve, Fig. 1, was determined has, within the limits of the curve, any efficiency between 2 and 7 watts per candle. Thus, by simply changing the candle-power of the lamp, we can operate it at any efficiency we choose, and get as much or as little light per watt as we choose.

In commercial practice the candle power of lamps is always marked on them and their efficiency at this candle-power is stated; but even this is not a proper index to the value of the lamp or to its proper efficiency. Experience has shown that lamps are almost universally run above their normal rating; lamps rated at 4.5 watts per candle are usually run at

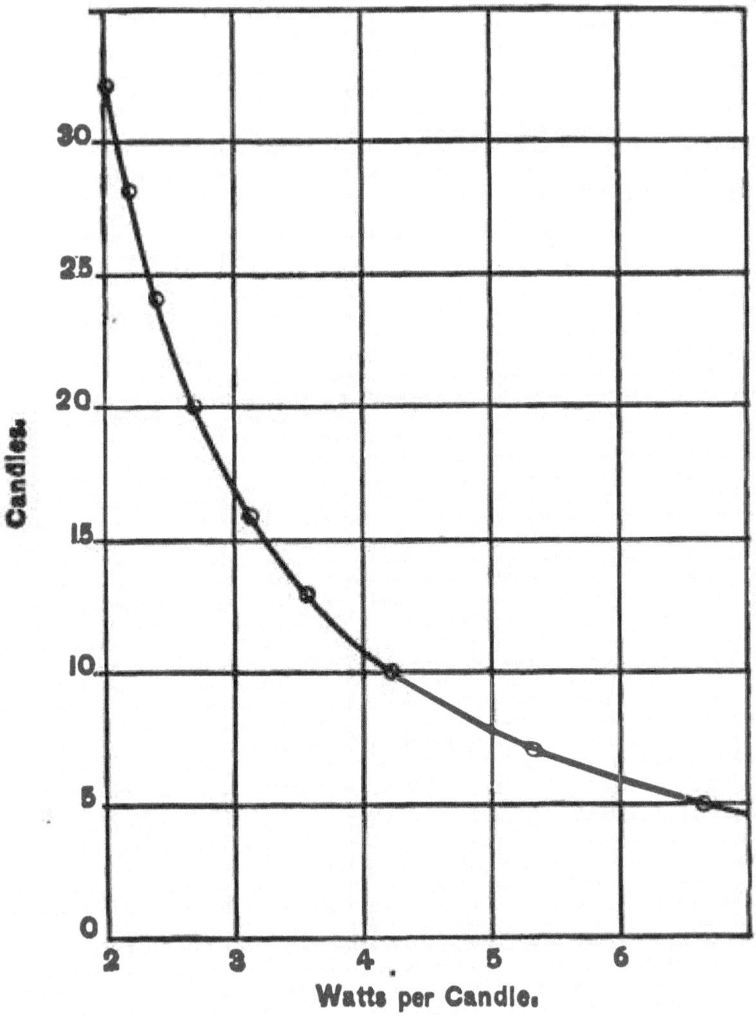

Fig. 1.

from 3.5 to 4 watts per candle, and in order to make lamps that will stand the strain of being run above their rated capacity, it is necessary to rate them considerably below the efficiency at which they will give the best results under ordinary circumstances.

Lamps have been made and sold in England which have a very high rated efficiency, but parties buying these lamps are told that they will get very much more satisfactory results if they run the lamp *below their rated capacity*. So we see that some lamps are rated above their capacity and some are rated below, and the rated efficiency of a lamp is not always the best efficiency at which to run it. How then, are we to determine the efficiency at which a lamp will give the best results? It is this question which I will attempt to answer.

The term "maximum efficiency" of a lamp, as used in the title of this paper, does not mean the highest efficiency at which a lamp can be operated, but the efficiency at which the best results are produced by the lamp: or more accurately, *the efficiency at which the cost of operating the lamp is a minimum.*

Taken in this latter sense, the maximum efficiency of a lamp is not its highest efficiency. As we increase the candle-power of a lamp its efficiency increases; consequently, by running the lamp high enough we can make its efficiency so high that very little power is required to produce a given amount of light, and the cost of power to produce the light is very small. But, while the efficiency of the lamp increases, its life decreases, and if we run a lamp at too high an efficiency the saving in the cost of power is more than balanced by the increased cost of lamp renewals.

To determine the maximum efficiency for lamps under given conditions, we must determine the efficiency at which the sum of the costs of power and lamps is a minimum, and in order to do this we must know the rate of variation of the life of a lamp with its efficiency.

The curve, Fig. 2, shows this rate of variation. This curve is the result of very carefully conducted experiments

Curve showing lives of equally good lamps, burned at different efficiencies.

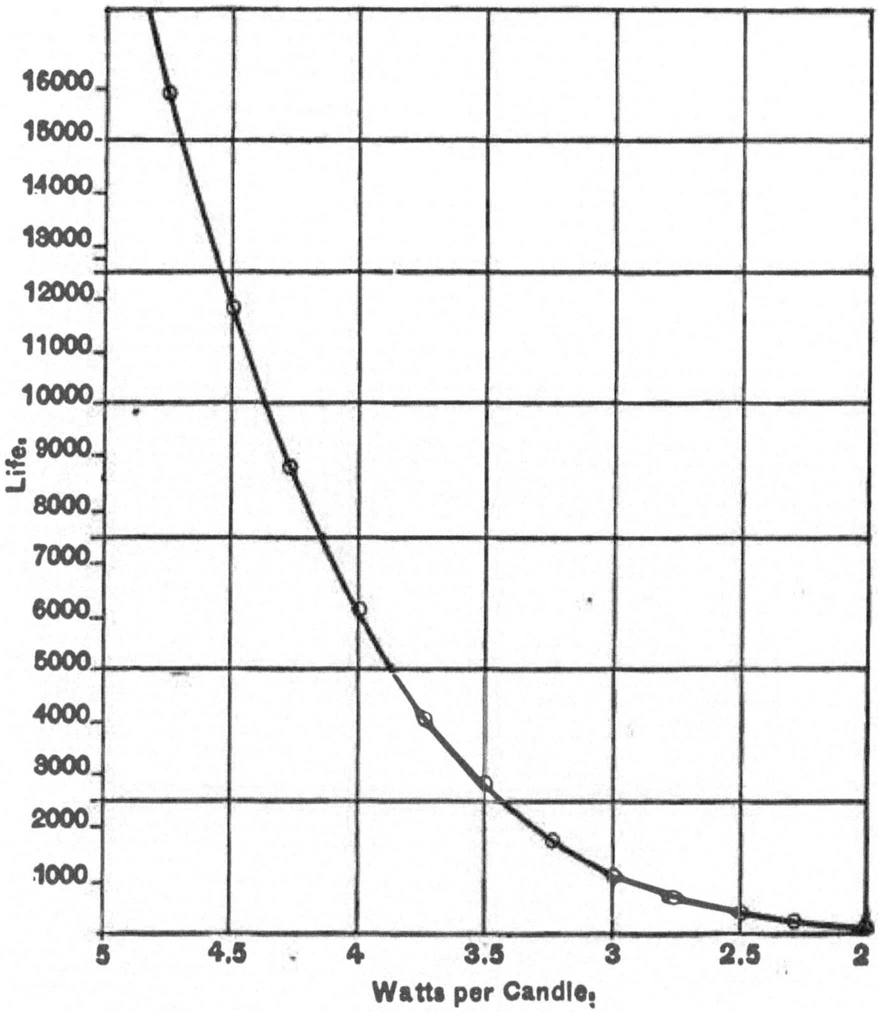

Fig. 2.

made by the Edison company. These

experiments extended over five years and consumed a very large number of lamps. Its accuracy when applied to Edison lamps is beyond question; but our experiments with lamps having an artificial surface on the carbon, or "flashed" lamps as they are called, show that their rate of variation of life and efficiency follows a different curve.

This curve does not apply to individual lamps. If we take two Edison lamps and burn them at different efficiencies, their lives for these efficiencies will probably not be such as indicated by the curve, nor will they be proportioned to these indicated lives. But if we take one hundred lamps and burn them at one efficiency, and another hundred equally good lamps and burn them at another efficiency, the average lives of the two sets will be proportional to the lives indicated by the curve for these two efficiencies.

In order to determine at what efficiency the cost of operating lamps of a given quality under given conditions is a minimum we must calculate what this cost is at different efficiencies. To do this we consider the total cost of operating the lamps to be made up of two parts, viz., the cost of the *current* and the cost of the *lamps*. The cost of the current is made up of every expense incurred in operating the lamps, including materials consumed, labor, taxes, insurance, rent and every other expense incurred in operating the plant, except the cost of lamps. The cost of the lamps is an item by itself, and is the amount which the lamp has cost when it is put in use, This is a natural division of the total cost of operating a plant, since to produce light by incandescence all that is necessary is a lamp and current to operate it.

If, in any case, we know the cost of the current required to operate the lamps, the cost of the lamp, the quality of the lamps—that is, the life they will give when burned at a given efficiency—and the rate of variation of their life with efficiency, we can then calculate at what efficiency the cost of operating the lamps is a minimum, and this I call the maximum efficiency of those lamps.

The following examples show what this maximum efficiency is under varying conditions of the cost of lamps, the cost of current and the quality of the lamps. The cost of the lamps I have varied between 25 cents and $1.00 each. The cost of current varies between 2.5 cents and 10 cents per h. p. per hour. The quality of the lamps varies between 300 hours life at 3 watts per candle, and 2,400 hours life at 3 watts per candle.

In each of the following cases I have calculated the cost of operating 100 16 c. p. lamps 1,000 hours, at each of the efficiencies comprised in the curve of total cost. These curves do not show the cost of running the same lamps at different efficiencies, but the cost of running equally good 16 candle lamps of the different efficiencies.

The first case we will consider is shown in the diagram Fig. 3. In this case the lamps are assumed to cost 85 cents each and to have a life of 600 hours at 3 watts per candle. The current is assumed to cost 10 cents per h. p. hour.

Lamps 85 cts. each—Life 600 hrs. at 3 Watts per Candle.
Current costs 10 cts. per HP. per hour.

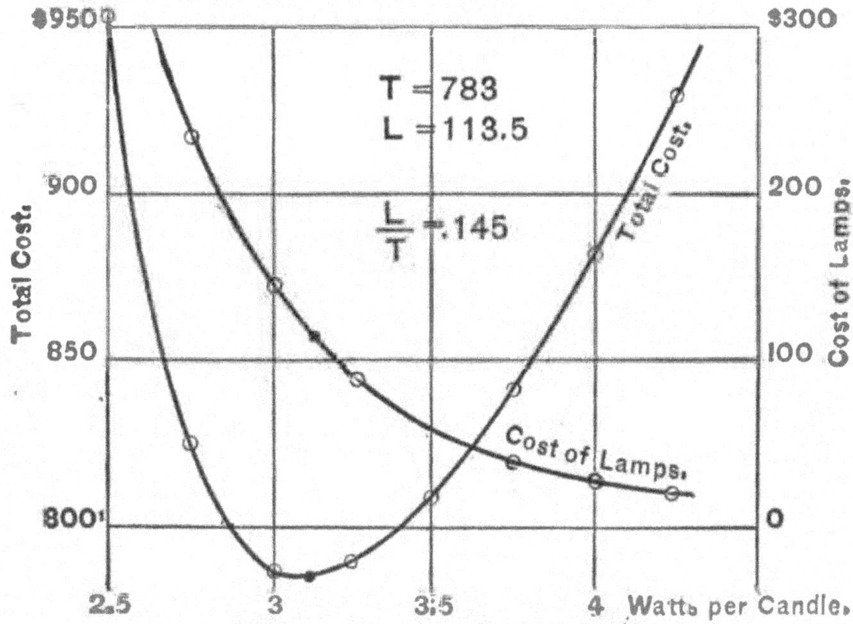

Minimum at 3.1 Watts per Candle.
Maximum Efficiency 3.1 Watts per Candle.

Fig. 8.

The cost of the current is determined from the following formula:

Current cost =

$$\frac{\text{w. p. c.} \times 16 \times 100 \times 1{,}000 \times \left\{\begin{array}{l}\text{cost of ct.}\\ \text{per h. p.}\\ \text{per hour}\end{array}\right.}{746}.$$

And the cost of lamps from this formula:

$$\text{Cost of lamps} = \frac{\text{cost of one lamp} \times 100 \times 1{,}000}{\text{Life at given efficiency}}$$

The curve marked *total cost* shows the total cost of running 100 16 c. p. lamps 1,000 hours, the efficiencies of the lamps varying between 2.5 and 4.25 watts per candle. The efficiencies are shown by the vertical lines, referring to the scale at bottom. The value of the total cost at any point of the curve is shown by the horizontal line through the point, referring to the scale at the left of the diagram.

The lowest point of the curve shows the point where the total cost is lowest. This is the minimum cost of operating these lamps under the given conditions. The mark at the lowest point of the curve shows this minimum cost to be $783, and a vertical line through this point to the scale at bottom of the diagram shows that this total cost is a minimum when lamps having an efficiency of 3.1 watts per candle are used.

Thus the maximum efficiency of these

Lamps $1.00 each—Life 600 hrs. at 3 Watts per Candle.
Current costs 10 cts. per HP, per hour.

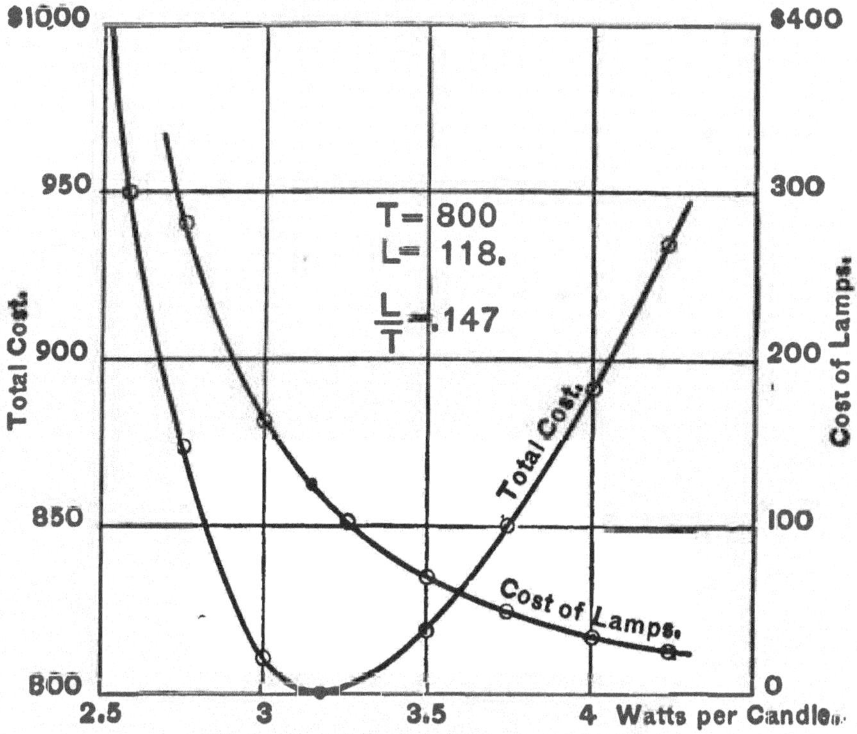

Minimum at 3.18 Watts per Candle.
∴ Maximum Efficiency 3.18 Watts per Candle.

Fig. 4.

lamps under the conditions assumed is 3.1 watts per candle.

The lamps considered in the case shown in Fig. 4 cost $1.00; all other conditions are the same as in the case shown in Fig. 3. This increases the total cost from \$783 to \$800, and necessitates using lamps of 3.18 watts per candle instead of 3.1,

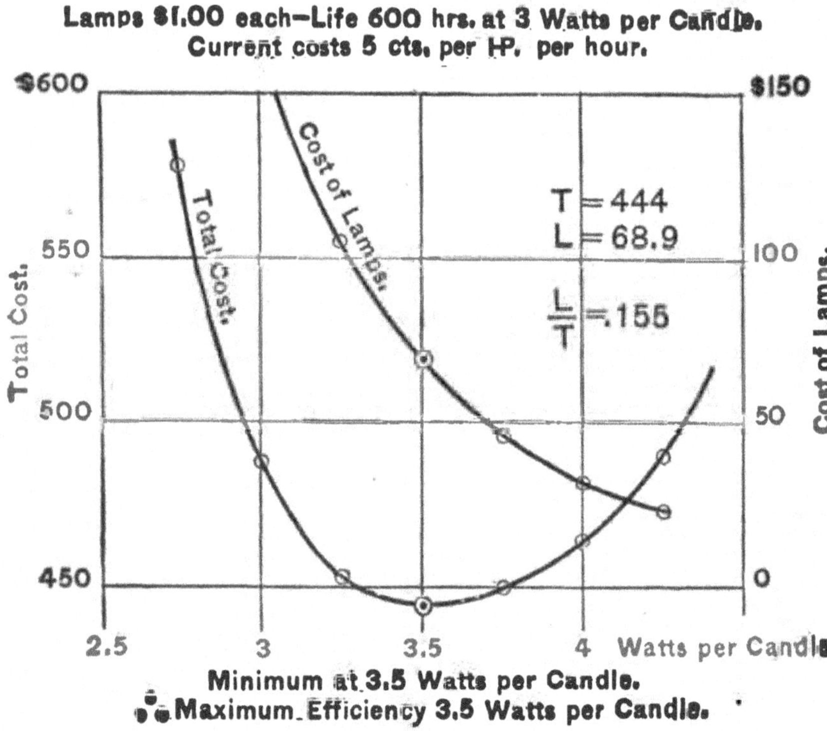

Fig. 5.

to make the cost of operating a minimum.

In the case shown in the diagram, Fig. 5, the current costs 5 cents per h. p. per hour; the other conditions are the same as in the case assumed in Fig. 4. The minimum total cost in this case is $444, and to make this cost a minimum we must use lamps having an efficiency of 3.5 watts per candle.

Lamps $1.00 each—Life 300 hrs. at 3 Watts per Candle.
Current costs 2.5 cts. per HP. per hour.

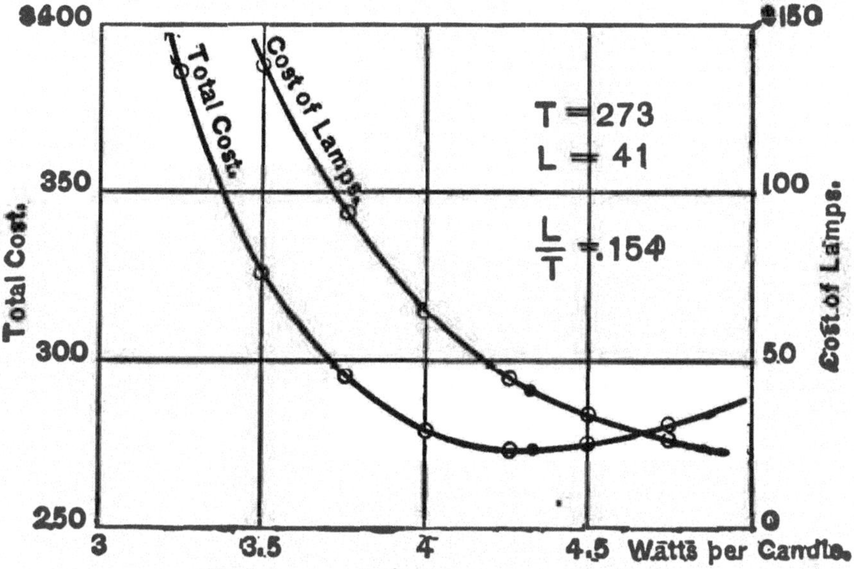

Minimum at 4.32 Watts per Candle.
& Maximum Efficiency 4.32 Watts per Candle.

Fig. 6.

In this case shown in Fig. 6 the lamps cost the same as in the last, but are only half as good; the current costs just half as much as in the last case. The minimum total cost in this case is $273, and the maximum efficiency of the lamps is 4.32 watts per candle.

In the case shown in Fig. 7 the lamps cost 50 cents and have a life of 1,200

Lamps 50 cts. each—Life 1200 hrs. at 3 Watts per Candle.
Current costs 10 cts. per HP. per hour.

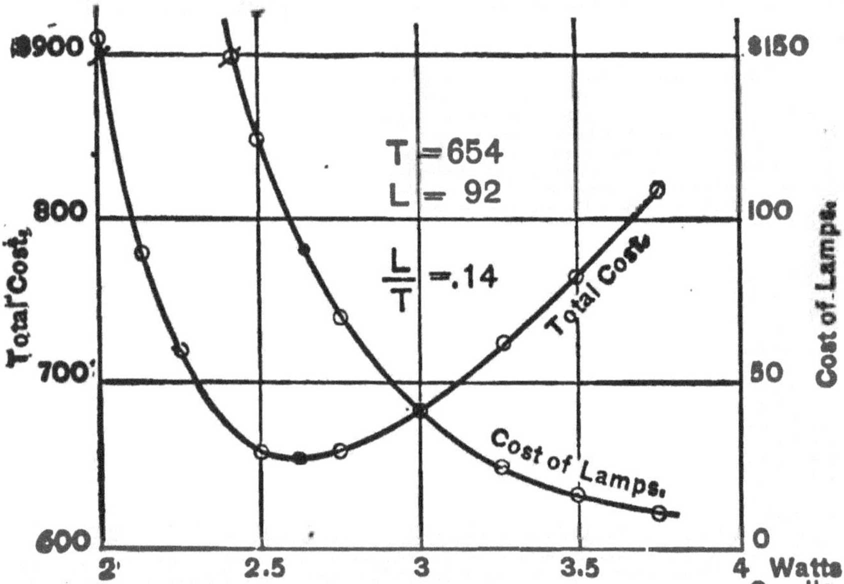

Minimum at 2.62 Watts per Candle.
Maximum Efficiency 2.62 Watts per Candle.

Fig. 7.

hours at 3 watts per candle. The current costs 10 cents per h. p. per hour. This is the cheapest and also the best lamp we have yet considered, but the current is expensive. In this case the minimum total cost is $654, and the maximum efficiency of the lamp is 2.62 watts per candle. In this case, Fig. 8, the current costs half as much as in the previous case, other conditions being the

Lamps 50 cts. each—Life 1200 hrs. at 3 Watts per Candle.
Current costs 5 cts. per H.P. per hour.

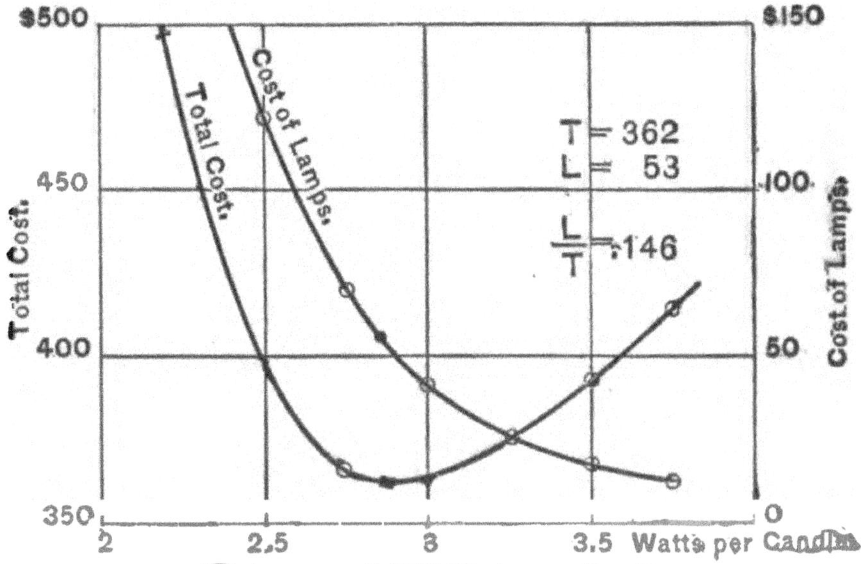

Minimum at 2.875 Watts per Candle.
• Maximum Efficiency 2.88 Watts per Candle.

Fig. 8.

same. The minimum total cost is reduced from $654 to $362. The maximum efficiency in this case is 2.88 watts per candle. In Fig. 9 the current costs twice as much as in the previous case and the lamps are only half as good. The minimum total cost is doubled, but the maxi-

Lamps 50 cts. each—Life 600 hrs. at 3 Watts per Candle.
Current costs 10 cts. per H.P. per hour.

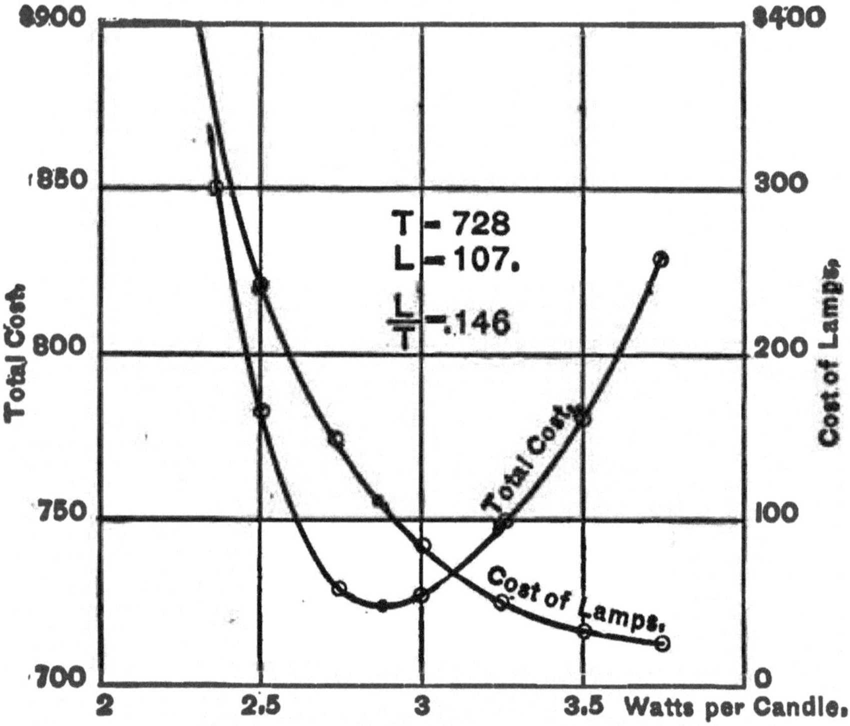

Minimum at 2.88 Watts per Candle.
∴ Maximum Efficiency 2.88 Watts per Candle.

Fig. 9.

mum efficiency is the same as in the previous case.

In this case, Fig. 10, the current costs one-half of that assumed in the previous case, other conditions being the same. The minimum total cost is $400, and the maximum efficiency 3.175 watts per candle.

The curve, Fig. 11, illustrates the case

Lamps 50 cts. each—Life 600 hrs. at 3 Watts per Candle.
Current costs 5 cts. per HP. per hour.

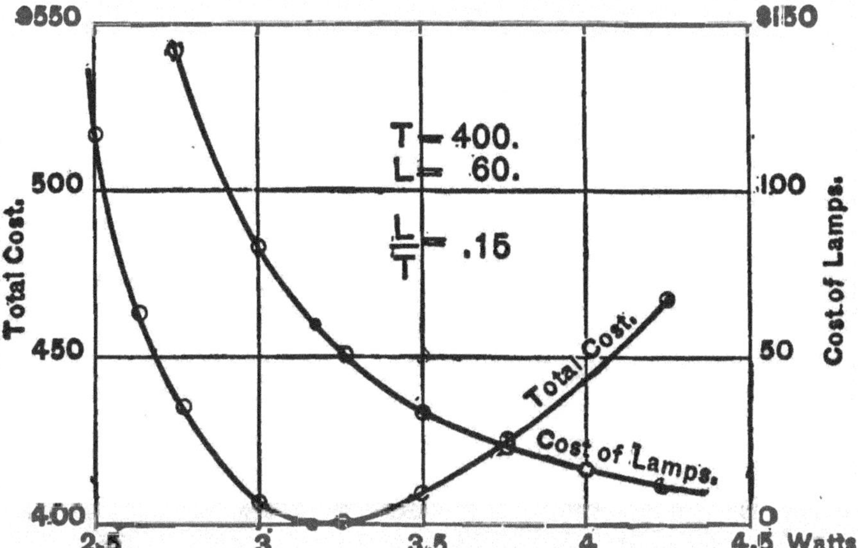

Minimum at 3.175 Watts per Candle.
Maximum Efficiency 3.175 Watts per Candle.

Fig. 10.

of very cheap and very good lamps, with moderate cost of current. The minimum total cost is low, $294, while the lamps are run at the high efficiency of 2.38 watts per candle.

In the case shown in Fig. 12, the cost and quality of the lamps are the same as in the previous case, but the current costs twice as much. This increases the mini-

Lamps 25 cts. each—Life 2400 hrs. at 3 Watts per Candle.
Current costs 5 cts. per HP. per hour.

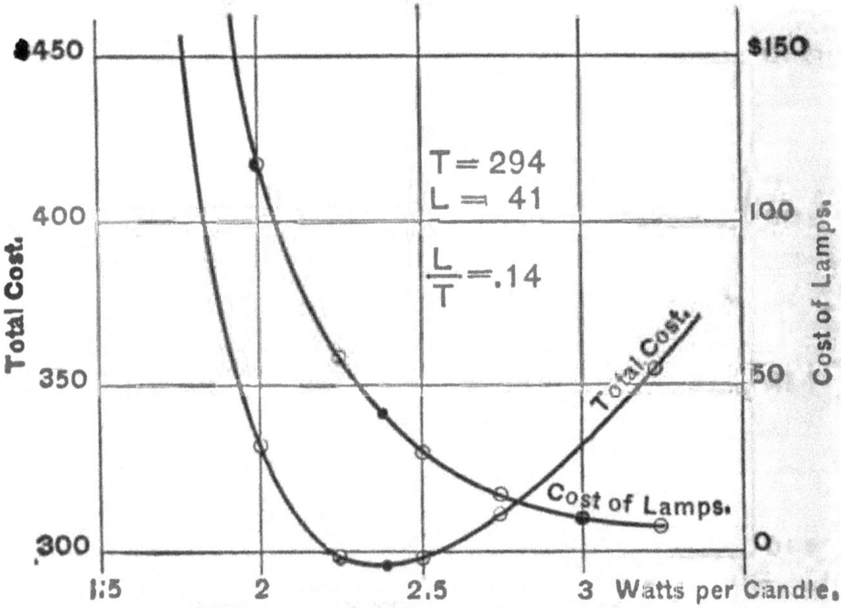

Minimum at 2.38 Watts per Candle.
Maximum Efficiency 2.38 Watts per Candle.

Fig. 11.

mum total cost from $294 to $535, and raises the maximum efficiency to 2.14 watts per candle.

In the case, Fig. 13, the cost of lamps and the cost of current are the same as in the case shown in Fig. 11, but the lamps are only half as good. The minimum total cost is increased from $294 to $327, and the maximum efficiency is reduced from 2.38 to 2.62 watts per candle.

Lamps 25 cts. each—Life 2400 hrs. at 3 Watts per Candle.
Current costs 10 cts. per H.P. per hour.

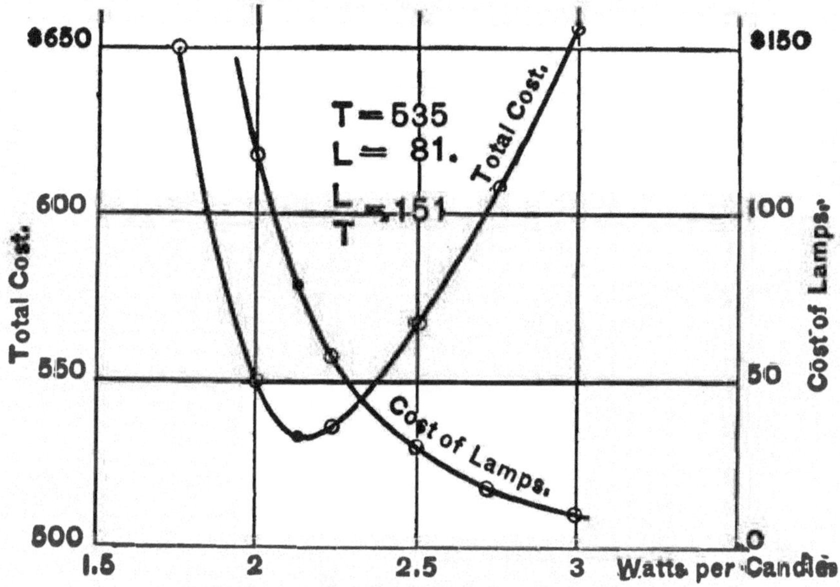

Minimum at 2.14 Watts per Candle.
Maximum Efficiency 2.14 Watts per Candle.

Fig. 12.

The first and plainest inference drawn from these curves is that the maximum efficiency of any given lamp is not a fixed one, but varies with conditions outside the lamp itself. Identical lamps operated under different conditions of cost of current must be burned at different efficiencies to make the cost of operation a

Lamps 25 cts. each—Life 1200 hrs. at 3 Watts per Candle.
Current costs .5 cts. per HP. per hour.

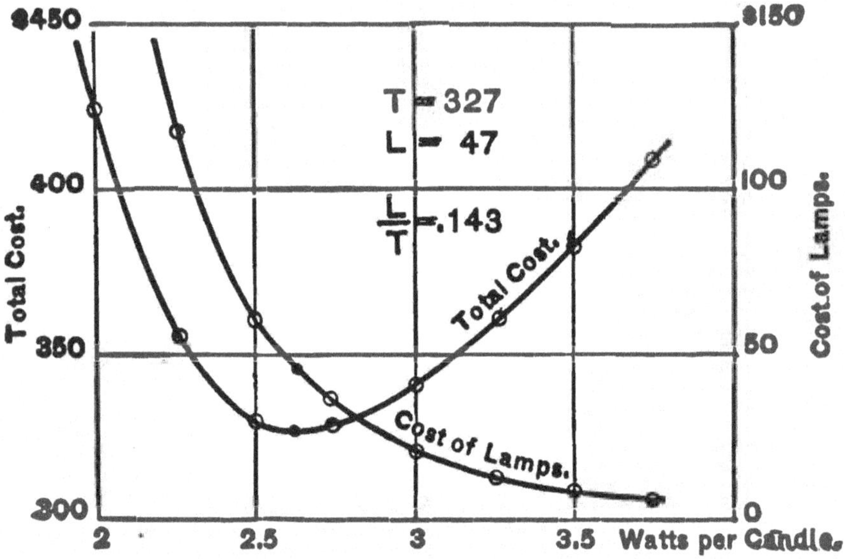

Minimum at 2.62 Watts per Candle.
Maximum Efficiency 2.62 Watts per Candle.

Fig. 13.

minimum for the production of a given amount of light. In order to determine the maximum efficiency of lamps, therefore, we must know the quality of the lamps referred to some standard, the cost of the lamps to the consumer, and the cost of current under the actual conditions existing at the place where the lamps are to be used.

In the eleven cases shown in this paper the lamp of highest efficiency is obtained in the case shown in Fig. 12. This is a case where the lamps are very cheap and very good, and current is very expensive. The lamp having the lowest efficiency is obtained in the case shown in Fig. 6, in which the lamps are poor and high-priced and the current is very cheap.

There is a marked difference in the sharpness of these curves at the minimum points. An inspection will show that the sharpness of the bend in these curves depends upon the cost of the current, the curves in which the current costs 10 cents per h. p. per hour being the sharpest. Those, in which the current costs 5 cents, are next, and the one, Fig. 6, in which the current costsonly 2½ cents per h. p. per hour, is very flat at the bottom or minimum point. In this comparison Fig. 7 is not considered, as it drawn on a different scale from the others, and is not as sharp as it should be.

This indicates that the more expensive the power is, the more carefully must the lamp efficiencies be chosen. In Fig. 12, which shows the sharpest curve, a very slight variation in the efficiencies of the lamps makes a very great change in the total cost of operation.

In the case shown in Fig. 6, in which the current is very cheap, we find that a very considerable change in the efficiency of the lamps used makes very little difference in the total cost of operation.

On each of the eleven diagrams two curves are drawn, one showing the total cost of operation, and the other showing the cost of lamps. On each of the curves showing the cost of lamps, a point is marked which indicates the cost of the lamps when the total cost is a minimum. The letter T marked on each of the figures denotes the minimum total cost of operating the lamps under the given conditions. The letter L denotes the cost of the lamps when the total cost is a minimum; and the expression $\frac{L}{T}$ denotes the ratio of the cost of lamps when the total cost is a minimum, to the minimum total cost. An examination of all these curves shows that while the minimum total cost varies with each of the three quantities—price of lamps, quality of lamps, and cost of current—nevertheless, *the total cost is always a minimum when the cost of lamps is about* 14.5 *or* 15 *per cent. of the total cost of operation.*

This figure varies somewhat in the different examples considered, but the variation seems to follow no law. In Figs. 6 and 12, which show the highest and lowest efficiency lamps, the figures are 15.1 per cent. and 15.4 per cent. respectively.

The steepness of the curve showing the cost of lamps, and the difficulty of determining the exact minimum point of a curve which has been drawn by inaccurate methods, makes it difficult to get the cost of lamps accurately when total cost is a minimum. These curves and values are given just as they were determined, and no effort has been made to bring the results into closer agreement, as could readily have been done. I consider the variation shown by these curves to indicate closely enough that this ratio of cost of lamps to total cost at the minimum point is nearly, if not quite, constant and that its value is between .145 and .15.

This establishes a very simple law for determining whether or not lamps are being operated at their maximum efficiency; for if they are, the lamp bills will be about 15 per cent. of the total operating expenses of the plant. If the lamp bills are more than 15 per cent. of the total operating expenses the lamps are being burned above their maximum efficiency, and lower efficiency lamps should be obtained. If, on the other hand, the lamp bills are less than 15 per cent. of the total expenses, the lamps are being burned below their maximum efficiency, and higher efficiency lamps would reduce the cost of operating the plant. Where fuel is high priced, or where other causes operate to make the cost of generating current high, it is specially important to use lamps of the maximum efficiency, for we have seen from the above curves that where the cost of current is high, the use of lamps whose efficiency is a little above or below the maximum efficiency attainable under the conditions of operation, makes a very marked increase in the operating expenses of the plant.

If in any plant the lamp bills are only 10 per cent. of the total expenses, then *increasing the efficiency of the lamps by increasing their candle-power does not diminish the total cost of operating the plant.* In order to reduce the total cost, *the lamps must be replaced by lamps of the same candle-power but of higher efficiency.* If the efficiency of the lamps is increased by raising their candle-power, the cost of operating the plant *per unit of light produced* is reduced, but the total cost is increased. A plant which is paying less than 15 per cent. of its total expenses for lamps, and which brings the lamp bills up to 15 per cent. by increasing the candle-power of the lamps, does not decrease the cost per lamp of operating the plant, but *does* decrease the cost per candle of light furnished. If they are paid for the increased light given by the lamps, the efficiency of the plant is made a maximum for the existing conditions; but if they are not paid for the additional light furnished, the efficiency of the plant is reduced.

This law enables any one operating an incandescent lamp plant to determine whether or not he is using the most suitable lamps for his plant. If the conditions of operation of a new plant are all known, and the quality of the lamps made by any lamp maker is known, we can determine before starting what is the most economical lamp to use.

www.ingramcontent.com/pod-product-compliance
Lightning Source LLC
LaVergne TN
LVHW050915210926
848482LV00012B/446